Advanced Manufacturing for Optical Fibers and Integrated Photonic Devices

Advanced Manufacturing for Optical Fibers and Integrated Photonic Devices

ABDUL AL-AZZAWI

ALGONQUIN COLLEGE, OTTAWA, ONTARIO, CANADA

CRC Press
Taylor & Francis Group
Boca Raton London New York

CRC Press is an imprint of the
Taylor & Francis Group, an **informa** business

CRC Press
Taylor & Francis Group
6000 Broken Sound Parkway NW, Suite 300
Boca Raton, FL 33487-2742

Printed on acid-free paper
Version Date: 20150630

International Standard Book Number-13: 978-1-4987-2945-1 (Hardback)

Visit the Taylor & Francis Web site at
http://www.taylorandfrancis.com

and the CRC Press Web site at
http://www.crcpress.com

Contents

Preface

We live in a world facing the challenge of developing technology that achieves simplicity, reliability, and cost savings. Manufacturing techniques are always under development. Some companies describe any advancement in manufacturing development as high-technology achievement. High manufacturing technology is related to very sophisticated manufacturing processes. Fiber optics, semiconductors, and laser products are all associated with high manufacturing technologies.

A unique approach is taken in this book to present new high manufacturing technologies and their applications in the high-tech sector. This book covers the theoretical principles and industrial practice of high manufacturing technology in a way that is suitable for students, engineers, professionals, and professors. Some chapters are presented in two parts: theoretical and practical. The theoretical part has adequate materials to cover all aspects of the subject. In the practical part, some chapters present the requirements for the theoretical concepts using both simple and advanced materials that are easy to understand. In this way, readers will learn and gain practical experience in manufacturing technologies. This will assist the readers in applying practical knowledge to real-world advancements. The step-by-step approach and technical illustrations in this book will guide readers through each chapter. The practical materials are designed for the application of high manufacturing technologies.

This book is written with simple language and gives adequate information and instruction to enable readers to achieve maximum comprehension. An effort has been made to use the international system of units (SI) throughout the book. The organization of the chapters is designed to provide a solid foundation for contemporary technology. This book abounds in theoretical and practical displays and is an effective teaching tool, helpful to both professors and students. Clear figures and readable tables are presented in this book to improve the comprehension of concepts.

The book is structured in 17 chapters as listed below:

- Chapter 1 covers workshop principles.
- Chapters 2 through 4 cover production lines, manufacturing processes, and clean rooms, which are needed for manufacturing products.
- Chapters 5 through 8 present high manufacturing technology for fiber optic cables, advanced fiber optic cables, installation, and fiber optic connectors, as well as passive and active fiber optic devices.
- Chapters 9 through 12 illustrate continuous improvement in a production line. 5Ss are one of the most important means and practices to reduce any type of waste in a production line. These chapters also present management's and employees' responsibilities in supporting production.
- Chapter 13 covers lean manufacturing processes for a production line.

- Chapter 14 covers product improvement. This chapter explains the outcome of improvement of the production line and all the steps that are taken to achieve the goal.
- Chapter 15 presents management supports in the professional manufacturing environment.
- Chapter 16 outlines the internal and external auditing of any production line. This chapter also highlights the ISO 9000 information and the advantages to auditing a production company.
- Chapter 17 presents high-tech manufacturing safety for a manufacturing workplace.
- The last part of the book includes a complete glossary of terms, references, and an index.

The book includes 168 figures and 8 tables. The book was developed using my extensive theoretical and practical experience with high manufacturing technologies in industry.

Abdul Al-Azzawi
Algonquin College
Ottawa, Ontario, Canada

Acknowledgments

I thank all of my family, colleagues, and friends who helped me complete this book.

In particular, I thank Mietek Slocinski for his, time, energy, fruitful discussion, and continuous support during the past few years.

I extend my thanks to Monica Havelock, Eng., and Mazin Altoumah, Arch. Eng., for their contribution of reviewing, proposing, and editing this book.

Thanks to my daughter, Abeer, and my son, Abaida, for their help in reviewing chapters and creating drawings and figures.

I express my great gratitude to my colleagues Prof. Devon Galway, Prof. Jim Catton, Prof. Robert Willbond, and Krysta Brennan, English editor, for their comments and feedback in reviewing some chapters.

I thank Lamyaa Shlah and J.P. Rouleau for their support during the time I wrote this book.

Abdul Al-Azzawi
Algonquin College
Ottawa, Ontario, Canada

About the Author

Dr. Abdul R. Al-Azzawi graduated from the University of Strathclyde in Glasgow, Scotland, UK. He has worked in the photonics manufacturing industry and as a professor at Algonquin College, Ontario, Canada. At Algonquin College, he is a professor, coordinator, and researcher in the Photonics Engineering Technology and Photonics and Laser Technology Programs. He is a professor teaching in the Bachelor Photonics and Laser Technology Program joint program between Algonquin College and Carleton University. He has published six books, many papers, and has participated in many workshops and conferences around the world. His special areas of interest are optic/optical fiber devices, fiber optic lighting, and fiber optic devices. He is a member of the Professional Photonics Society in Canada and Senior Member of the Institute of Electrical and Electronics Engineers (SMIEEE). He is the recipient of the National Institute for Staff and Organizational Development (NISOD) Excellence Award from the University of Texas at Austin in 2005.

While employed as a photonics engineer in the photonics industry, he designed new production lines, modified products, wrote manufacturing processes, designed new jigs, and planned new products for manufacturing.

1 Workshops

1.1 INTRODUCTION

At the beginning of the Industrial Revolution, workshops were developed for mass production. A workshop can be a small room, a big bungalow, or a building that provides the facilities for a manufacturing floor, tools, machinery, and systems. The facilities may be required for the manufacture or repair of manufactured goods. Workshops can be developed for a big mass production that would lead to a large factory. Workshops were developed in the last century using high manufacturing techniques for high-technology devices and systems productions. Workshops can be used for the production of goods or for social and other required activities in science, technology, planning, and teaching.

1.2 STANDARD WORKSHOPS

Standard workshops are the most used space for manufacturing goods in small and large quantities. The workshops can be for one type of manufacturing product or multiple purposes. Workshops are used for mechanical, wood, and metal work products. Workshops are classified as follows:

- Academic workshop
- Acting workshop
- Agricultural workshop
- Air-space workshop
- Building woodshop
- Chemical workshop

- E- or online workshop
- Educational workshop
- Electrical workshop
- Electronic workshop
- Mechanical workshop
- Multiple production workshop
- Music workshop
- Office workshop
- Photography workshop
- Photonics workshop
- Production workshop
- Series workshop
- Sheltered workshop
- Sociology workshop
- Songwriter workshop
- Telecommunication workshop
- Training workshop
- Etc.

1.3 HIGH-TECH WORKSHOPS

Standard workstations are used by employees during the entire shift. The station is used for the productivity, health, safety, and comfort of employees, as well as for promoting effective interaction among employees, technologies, and the environment in which both must operate. Figure 1.1 shows a standard workstation. The station should be accommodating so that personnel may interface effectively with tools and devices. Tools, jigs, and devices should be sized to fit the station and each individual

FIGURE 1.1 Standard workstations.

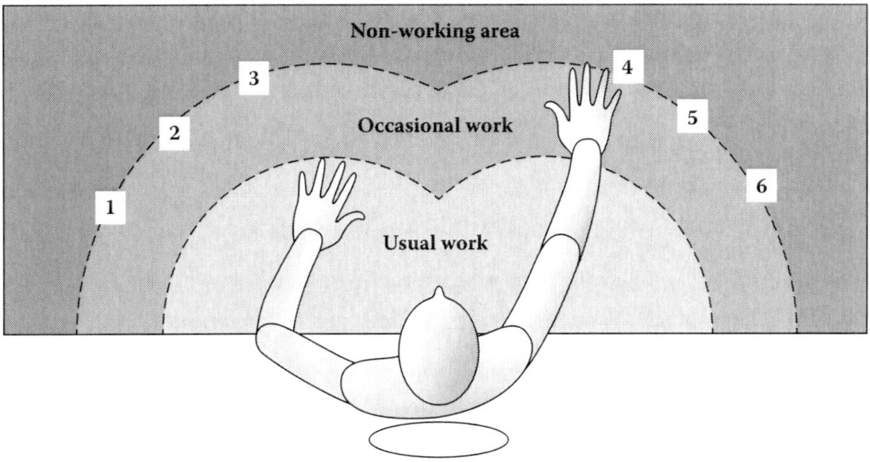

FIGURE 1.2 A standard workstation with six assembly parts.

user. The station should be designed to facilitate task performance and minimize fatigue and injury by fitting tools and devices to the body size, strength, and range of motion of the user. Some stations are equipped with shelves, one or double fluorescent lights, a true earth connection for electrostatic discharge (ESD) devices, electric outlet connections, and an air pressure supply gun. There are many sizes of standard stations. The most preferable size is 75 × 150 cm (2.5 × 5 ft), as shown in Figure 1.1.

Figure 1.2 shows an employee working on a single assembly standard workstation. The manual part dispensers for the six different single assembly parts are arranged in semicircular form around the assembly position. The numbering of the manual part dispensers corresponds to the single part numbers as shown in Figure 1.2. The mean distances for reaching and collecting between the work piece position and the individual gripping positions of the six parts should be in a comfortable reaching position to the employee, according to health and safety regulations. Each workstation type has its standard regulations for working distances for reaching, gripping, collection, assembly, and release.

The station work surface should be adjusted so that it is elbow height. If it cannot be adjusted, the height of the chair can be adjusted. If necessary, a footrest can be used. Furthermore, materials that are used frequently should be located within easy reach. Other materials should be placed just outside this easy-to-reach zone. It is also a good idea to put some materials completely out of reach. This will force the employee to get out of the chair and move around, promoting blood circulation and therefore reducing the stress on the employee's body.

1.4 RAPID MANUFACTURING WORKSHOPS

The rapid development of workshops is as a result of using new manufacturing techniques, which produce products with low cost and according to a customer's standards. Current manufacturing techniques are mostly controlled by automation setups. The manufacturing processes are mostly controlled by a computer using a mathematical code. Rapid workshops done in parallel batch production can provide an advantage in speed and lower cost compared to alternative manufacturing techniques, such as plastic injection molding or die casting. Rapid workshops can be used in the production of custom design parts, general consumption parts, standard production lines, large products, and series production.

1.5 ADDITIVE MANUFACTURING WORKSHOPS

Additive manufacturing workshops are the process of making a product by adding layers in a relatively efficient way, such that there is little waste or reduction of materials and the floor area is used efficiently. In this way, the addition of a manufacturing process can be implemented to an automated production line in a rapid workshop. The most popular examples are the production of inkjet printing, aerosol jet printing of electronic circuits, and 3D printing.

1.6 AN EXAMPLE OF WORKSHOP MODIFICATIONS

As described in this book, a workshop that is a room, an area, or a small establishment has many types of tools and machines where manual or light industrial work can be done. Workshops can be used for many types of manufacturing products. Workshops also have a combination of tools and machines for multiproduct design selections. For example, examine the combination of wood and metal work machines shown in Figure 1.3. This figure shows the machines for wood and metal works are distributed randomly around the floor. This example was a project completed by the author.

As shown in Figure 1.3, the workshop uses tools, jigs, and machines for different wood and metal products. Some of them are shown in Figure 1.4. This workshop operates in the traditional way of using tools, jigs, and machines. In this way, the workshop can produce many product designs.

Machines provide high power for small jobs in general design products. The conditions of the machines consume energy and have high operating costs. Consequently, the cost will add on to the price of the products. Figure 1.3 also shows that a combination of wood and metal work machines is a fire hazard due to flying sparks from metal cutting and grinders. The machines are also not arranged in subsequent manufacturing steps (one step followed by the next) to easily complete a product.

Let us audit the workshop setup and performance. The following can be highlighted:

- Machines, service machines, and tools are distributed randomly
- High power consumption machines
- Large size machines
- Missing manufacturing processes and file documentations
- Weak lighting distribution
- Old air-conditioning systems and uniform air distribution in the workshop
- Inappropriate air section fan locations
- Low employee morale
- Bad air quality generated from some of the machines, spray painting, and cleaning materials

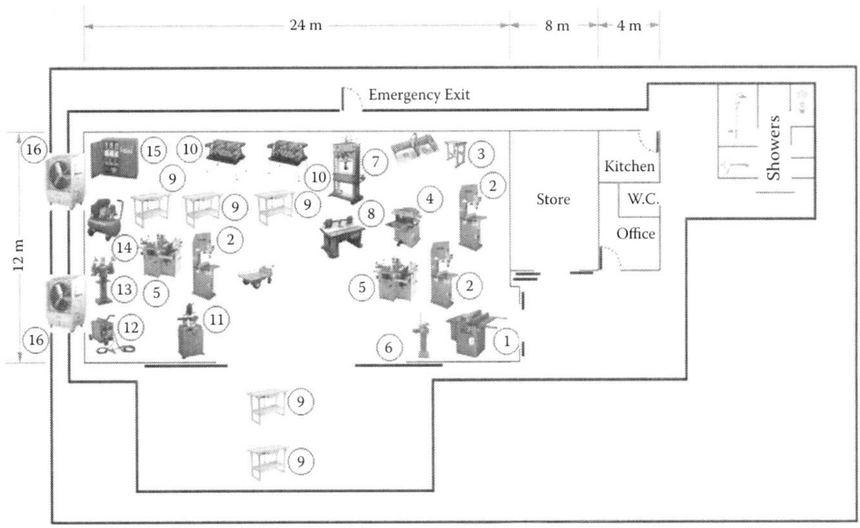

1 – Table Saw	7 – Manual Hydraulic Press	13 – Pedestal Grinder
2 – Band Saw	8 – Router	14 – Spray Gun with Compressor
3 – Manual Sharpener	9 – Working Bench	15 – Paint Storage Cabinet
4 – Thicknesser	10 – Portable Working Table	16 – Air Cooler
5 – Compound Machine	11 – Abrasive Circular Saw for Metal	
6 – Band Saw Sharpener	12 – Welding Machine	

FIGURE 1.3 Wood and metal workshop.

- No inventory lists for the items and products
- Bad production storage locations
- Lack of employee training
- Poor product scheduling
- Poor employee scheduling
- No regular maintenance schedule plan
- Lack of updated drawings, procedures, and documentations
- Missing some fire extinguishers
- Missing items in the first aid box
- No emergency door and directive signs
- Missing pages from the Workplace Hazardous Materials Information System (WHMIS) manual
- Some worn tools, jigs, and cutter blades
- No internal and external auditing
- Extra cost added on the products

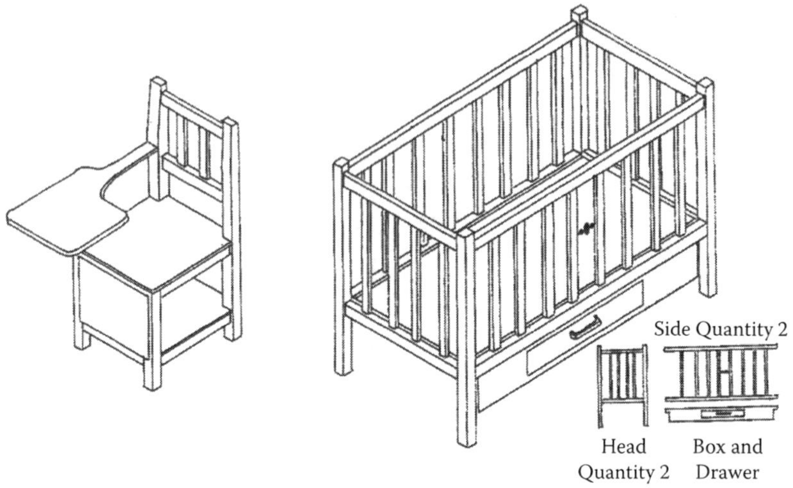

FIGURE 1.4 Furniture products.

There are many easy modifications that can be carried out on this workshop. Let us make the following changes and observe the benefits:

- Use the floor efficiently and rearrange the machines, service machines, and jigs so that machine arrangement looks like a production line.
- Use handling systems like roller reeling between the machines in order to become a production line. In this case the handling time is reduced drastically for moving items between assembly stations.
- Add dollies.
- Use suitable tools and devices for any job.
- Use appropriate motors for each machine to reduce electric consumption.
- Update and document the manufacturing processes and file the documentations.
- Choose, install, and maintain an energy-efficient lighting system.
- Use energy-efficient air conditioning units.
- Install and maintain exhaust fans.
- Implement an employee training plan.
- Use automatic product scheduling.
- Use automatic employee scheduling.
- Establish a regular maintenance schedule plan.
- Update drawings, procedures, and documentations regularly.
- Assign a safety committee from management and employees and check each fire extinguisher regularly.
- Assign one member of the safety committee to take care of the first aid box.
- Add emergency door in the back of the building.
- Safety officer will take the responsibility to check each single item and check reports related to safety issues in the building. The WHMIS manual

and material safety data sheet (MSDS) papers should be kept in an accessible folder for all employees.

- Replace the worn tools, jigs, and cutter blades. With good tools you will get a good product.

Figure 1.5 shows a few changes implemented to improve the workshop. Some very necessary changes, even though it costs extra money, are adopted. The extra money will pay off shortly, as it's an investment to the benefit of the workshop. There are further changes that could be implemented, but they can be costly, such as replacing some machines and jigs. Let us audit the workshop again after all the simple changes listed above are adopted. The energy consumption is reduced by 26%. The floor area is reduced to accommodate the new machine arrangements by 30%. By adding an air-conditioning unit and exhaust fans, the air quality improves drastically; all employees will feel very comfortable working long hours without any building syndrome (getting health effects, such as fatigue, red eye, back pain, etc.).

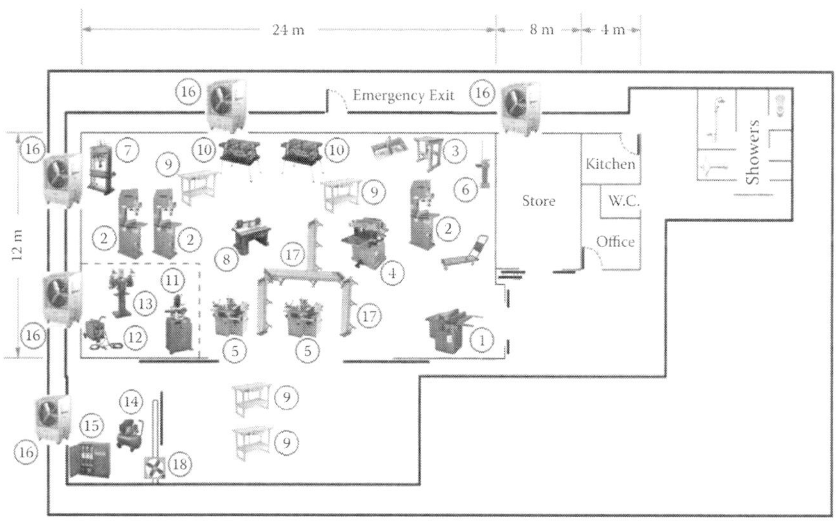

1 – Table Saw	7 – Manual Hydraulic Press	13 – Pedestal Grinder
2 – Band Saw	8 – Router	14 – Spray Gun with Compressor
3 – Manual Sharpener	9 – Working Bench	15 – Paint Storage Cabinet
4 – Thicknesser	10 – Portable Working Table	16 – Air Cooler
5 – Compound Machine	11 – Abrasive Circular Saw for Metal	17 – Roller Conveyor Sections
6 – Band Saw Sharpener	12 – Welding Machine	18 – Exhaust Fan

FIGURE 1.5 A new modified design of the wood and metal workshop.

The changes have a positive effect on the employees' morale. Because the employee training plan was scheduled for everyone, every employee is now involved directly in the manufacturing processes. The use of wireless communication systems between employees and management saves considerable time.

The use of simple automation in a few machines and in the programmable cutters and drills allows for very smooth cuts and therefore reduces time. Also, the use of computerized employee and product scheduling facilitates work orders and quick on-time delivery. Improved safety is very welcomed by all staff. On the top of it all, there is a 12% reduction in product and overhead costs. This reduction is reported as a profit to the company along with all other positive effects. This example proves that any workshop can be modified and updated for better product with low costs in order to be competitive and healthy in the market. It is also possible to invest money to update the manufacturing facilities using advanced technologies. Currently, automation is the best way to stay competitive in this market.

2 Production Lines

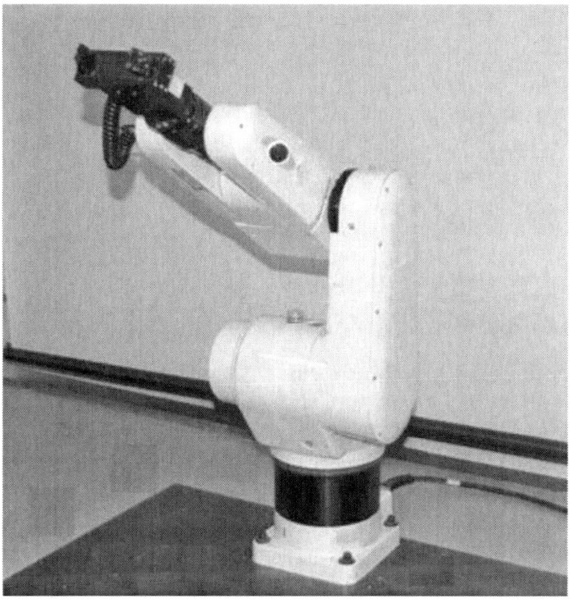

2.1 INTRODUCTION

Assembly or production lines can be defined as "an arrangement of workers, machines, tools, and equipment in which the product being assembled passes consecutively from first operation to next operation until completed." The production line can be one operation needed to complete a product, or multiple operations could go on, with more than 10 or even 20 operations to complete a product. Mechanical production methods require workers to perform a repetitive task on a product as it moves along on a conveyor belt or track. In photonics production, a product can be moved between stations using one type of tray as a holder and handler, while at the same time moving the product. An assembly line has the advantages of repeating the work and standardizing the product.

2.2 WHY BUILD PRODUCTION LINES?

The following essentials are encouraged prior to establishing a production line:

1. Consumer demand
2. Economic necessity (to reduce production costs)
3. Standardization of products
4. Controlled production

2.3 PRODUCING ECONOMICAL PRODUCTS

The following steps are important in order to create an economical production line:

1. Study the market and consumer demand.
2. Plan the production line.
3. Create the production line.
4. Create a research and development group.
5. Improve the products regularly.
6. Use alternative and high-quality materials.
7. Train and educate employees (management and technical personals).
8. Use statistical cost analysis and make products competitive.
9. Create short- and long-phase planning.

2.4 REQUIREMENTS TO BUILD AND DESIGN A PRODUCTION LINE

The following requirements are needed to build and design a production line:

1. Location (building site).
2. Area to accommodate the production line, storage, management, services, parking, etc. Sometimes a rented area is enough to accommodate a new production line. Sometimes a new site for a new or modified production line needs to be built.
3. Create a design for the floor plan to include the major elements that are required to build a production line in a building, such as:
   a. Space
   b. Doors (entrances/exits and emergency exits)
   c. Windows (fixed and operable)
   d. Pressurized air supply
   e. Water supply
   f. Electrical power supply (120, 220, and 380 V)
   g. Air-conditioning system
   h. Lighting systems (general lighting system and emergency lighting system)
   i. Storage space (at room temperature, heating and cooling spaces, and deep freeze)
   j. Shipping and receiving areas
   k. Offices (president, board meeting room, managers, engineers, technologists, technicians, operators, human resources, technical support, etc.)
   l. Meeting rooms
   m. Multiactivity area
   n. Cafeteria and kitchen facilities
   o. Toilet and bathroom facilities
   p. First aid and treatment rooms
   q. Mechanical, electrical, and hydro rooms
   r. Telecommunication and control rooms
   s. Parking area, and so on

4. Station types, such as model, class, type, special, electrostatic discharge (ESD), etc. (Consider the need for hoods or pressurized air supply.)
5. Jigs.
6. Ovens.
7. Systems (test, measurement, inspection, etc.).
8. Telephones and computers.
9. Management employees.
10. Technical employees (engineers, technologists, supervisors, technologists, technicians, operators, human resources, technical support, financial officers, planners, marketing, etc.).
11. Warehouse.
12. Many other requirements that can be added depending on the production lines and products.

2.5 CATEGORIES OF MANUFACTURING LINES

There are three main categories of manufacturing lines:

- Continuous (fixed) manufacturing lines. Most manufacturing lines are in this category. Workstations, ovens, some measurement and test devices, and tools are used to build a production line. Usually there is no need for jigs, special tools, or equipment that is specially designed for specific manufacturing processes. The classical manufacturing production lines for this category are cars, computers, and fiber optical devices.
- Flexible (multiuse) manufacturing lines. The most popular examples for this category are a machine shop, carpentry workshop, toolmaking workshop, etc.
- Batch manufacturing lines. In batch production, a production line is built for one design of one type of product in limited quantity, for a specific certain order at a certain time.

2.6 MANUFACTURING PROCESSES

There are six types of manufacturing processes:

1. A single-unit production job performed by one skilled person, as shown in Figure 2.1. This figure shows that manufacturing of a work piece is done by one skilled employee.
2. A single-unit production job performed by one skilled employee and unskilled employees.
3. A flexible production process (batch job and multiple uses).
4. A small arranged production process.
5. A partially automated production process.
6. A fully automated production process, as shown in Figure 2.2. This figure shows the manufacturing process in the production line using a robot to build work pieces.

FIGURE 2.1 A skilled employee building a work piece.

FIGURE 2.2 A fully automated production process.

2.7 TYPES OF ASSEMBLY LINES

Assembly lines are generally divided into the following types.

2.7.1 Single Production Lines

Single production lines are an industrial arrangement of machines, equipment, and workers for the continuous flow of work pieces in mass production operations. An assembly line is designed by determining the sequences of operations for the manufacture of each component as well as the final product. Figure 2.3 shows an example

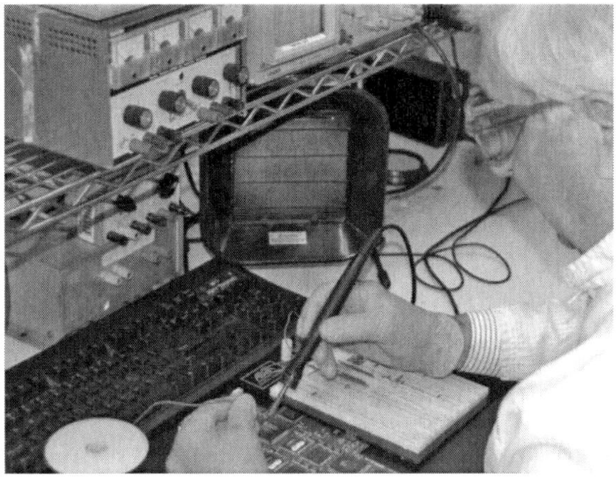

FIGURE 2.3 A single production line.

of an assembly station in a single production line. Each movement of material is made as simple and short as possible, with no cross-flow, bottleneck, or backtracking. Work assignments, the number of machines, and production rates are arranged so that all operations performed along the line are compatible. Single production lines are one of the mass production methods, which reduce lead cost along with increased quantity and quality. The requirements for mass production of a particular product include the existence of a market large enough to justify a big investment in building a production line. The floor plan of a mass production line ensures minimized material handling time, reduced manpower, the use of automated machines, repetitive steps, continuous flow of work spaces, and tools or jigs designed specifically for the tasks to be performed.

2.7.2 Parallel Production Lines

Two approaches in configuring parallel manufacturing lines are currently being used in industrial plants. These have been characterized as the Japanese approach of parallel independent cells of serial operations and the European approach of a serial line with each operation being duplicated in parallel. The European approach has a productivity advantage over the Japanese approach when considering machine failures within each operation. However, the European approach requires more material handling, which increases the configuration complexity and can reduce productivity. Using parallelism in bufferless production lines can improve productivity, with significant productivity gains achieved with crossover. However, including crossover in the line requires additional material-handling requirements that may reduce the availability of the system.

Many semicontinuous plants use several production lines and shared resources to make multiple products for various due dates. Short-term scheduling of such plants with sequence-dependent transitions, inventory costs, safety stock penalties, and backorder penalties is an advantage of the parallel production line. Figure 2.4 shows a parallel production line.

FIGURE 2.4 A parallel production line.

2.7.3 Feed in Production Lines

Feed in a production line is the structure for a complex assembly line. It begins as one main line with stations along it that are fed by lines running perpendicular to it. Each of these side lines is a feeding component for the finished product. The assembly line can be easily changed to handle changes in the product modifications or new product version. Examples of this type are upgraded computers and special car orders.

2.7.4 Cross-Production Lines

A cross-production line has two parallel running production lines that have the ability to cross over depending on the specifications of the product. Examples would be the production line of a computer and machines with different versions and upgrades.

2.7.5 Partially Automated Lines

A partially automated manufacturing line is similar to a fully automatic one. The difference is that in a partially automated manufacturing line, a worker uses a machine to aid in the assembly of the product. Machines can help in many areas from heavy lifting to precision drilling. An example of a partially automated assembly line is the production line that produces a product from an automated machine and is packed by a worker, as shown in Figure 2.5.

2.7.6 Fully Automated Lines

Fully automated assembly lines consist entirely of machines run by other machines and are used in continuous processes. The machines assemble the product from the beginning all the way to the end product that is ready for shipping, as shown

FIGURE 2.5 A partially automated line.

FIGURE 2.6 A fully automated production line.

in Figure 2.6. The whole manufacturing line is monitored by staff to ensure the machines are doing the proper job. Examples of fully automated assembly lines are an oil refinery where production never stops and industries such as petroleum refining and chemical manufacturing, as well as many modern automobile engine/body plants.

Most fully automated lines use fully automated machines and robots to perform manufacturing processes. Robots have the potential to revolutionize the manufacturing process, improving repeatability, cycle rate, reliability, and safety on the plant floor. This will lead to reducing manufacturing costs associated with labor, and time. The robots are mechanical systems that operate with software programs to perform the mechanical actions by moving the arms or the heads that carry a set of tools. The set of tools can be for cutting, drilling, taping, etc. There are many types of robots that are available for different manufacturing processes. Some applications of these robots are as follows:

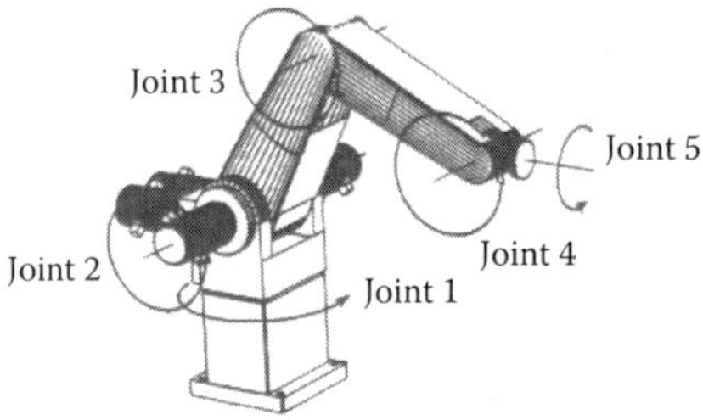

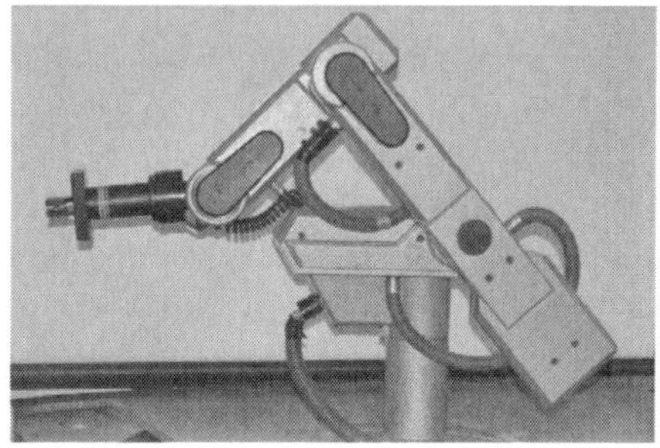

FIGURE 2.7 A multiaxis robot.

1. Multiaxis robots. Multiaxis robots are designed to bring high performance, precision, motion, and full multiaxis dexterity to clean room assembly, handling, testing, and packaging applications. They can be used in many production lines for manufacturing solar panels, disk drives, LCDs, and semiconductors, and in many other applications. These robots can be designed in two-, three-, four-, five-, and six-axis motions. Figure 2.7 shows a six-axis robot.
2. Vision-guided robots. Vision-guided robots are regular robots fitted with a camera or sensor that is used to provide positional information to the software so that the position is changed to adapt to the real position of the workplace or part being processed by the robot. The vision system is used to determine a new position and hence guide the robot to a new process. The major benefits of this type of robot are the reductions in labor costs and the increase in asset utilization due to the ability to run multiple part styles down the same production line. Figure 2.8 shows a vision-guided robot.
3. Picking robots. Picking robots are designed for specific jobs, such as handling primary food and packaged products. This type of robot is capable of picking products at a high speed of cycles per minute on a continuous basis while using visual line tracking. Depending on the application, these robots can be compact and can be mounted in a variety of positions in a production line. The picking robots can be used in many applications, such as machine tending, assembly, picking and packing, part washing, material removal, clean room operations, and testing and sampling.
4. Robot sensors. This type of robot is designed with an onboard control and sensing functions to enable seamless integration to robotic production work cells. This robot is used in laser welding, monitoring, and the weld inspection processes that can deliver high productivity and quality in any manufacturing process. The robot is also used for a variety of material handling applications, including material handling, machines load and unload, measurements, and palletizing and depalletizing. Utilization of these robots will play a significant role in the improvement of manufacturing efficiency.
5. Robot tools. Robot tools are designed to provide real-time, high-accuracy 3D data for the inspection of critical dimensions or to optimize quality

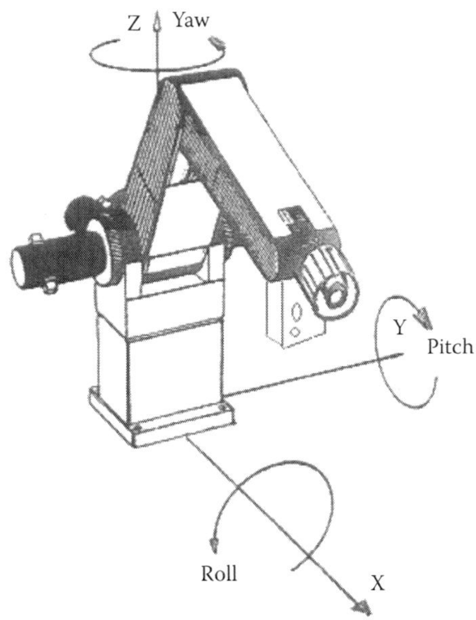

FIGURE 2.8 A vision-guided robot.

control in almost any manufacturing process and production line. This robot has a 3D laser scanning head and a high-performance 3D data acquisition and processing library for the creation of acquisition sequences and automated inspections. The system is mounted on the end of the multiaxis robot. The scan device converts the data to produce a highly accurate representation of the surface to be inspected. This robot can be used for quality control on the high-speed data rate.

6. Other robot types and applications. There are many robot types and applications that are used in different manufacturing processes. As mentioned above, the robots can be universal for doing more than one job and can be moved from one production line to another, such as the robots that are used in body car building. On the other hand, other robot types are designed for one specific job, for example, the robots used in chips manufacturing and integrated circuit wiring.

2.7.7 Pilot Production Lines

Pilot production lines can be designed for new product implementation. The new product comes from the R&D for trial production. Pilot lines have their own specific requirements. Many changes can be expected during the trial process. The main objective is to turn the pilot production line into a mass production line after calculating and designing the best manufacturing process to achieve mass production. Manufacturing, process, test, and quality control engineers can use familiar equipment, tools, and jigs backed by R&D development data and analysis.

2.8 STANDARD WORKSTATIONS

As explained in this book, there are many types of workstations that are used in conventional manufacturing lines. A standard workstation has a top working area of 75 × 150 cm (2.5 × 5 ft). These stations are equipped with shelves and lighting systems. They are also available in different sizes. Some of the stations are physically grounded (true earth), as is needed in manufacturing electrostatic-sensitive devices and pressure air supply for cleaning.

2.9 PRINCIPLES OF BUILDING SUCCESSFUL PRODUCTION LINES

A successful production line adopts the following:

- Reliability: Reliability is an essential element in any manufacturing line.
- Flexibility: Flexibility is another essential element in any manufacturing line. It covers a wide spectrum of manufacturing activities that include fabricating and assembly.
- Productivity: Productivity is an essential element in any manufacturing or nonmanufacturing activity, in order to stay competitive and increase productivity.
- Integration: Integration of line components so that they work together and in harmony is based on a set of rules identified by line integration.
- Immediacy: Immediacy is the speed with which the line can react to various changes.
- Economical: Achieving economic and competitive benefits.

2.10 CONSEQUENCES OF BUILDING UNSUCCESSFUL PRODUCTION LINES

The following are some of the factors contributing to the increased cost of products produced in an improperly designed production line. These are a very good recipe for failure.

- Poor planning and management
- Poor utilization of machines and tools
- Increased process time
- Untrained employees
- Poor standards and specifications
- Increased in-process inventories
- Poor quality controls
- Poor use of jigs, handling, storage, and shipping facilities
- Inefficient use of floor space
- Poor inventory management

2.11 CAUSES OF DAY-TO-DAY PROBLEMS

The following can cause day-to-day problems:

- Engineering changes
- Processing changes
- Machine breakdown
- Deterioration in tools and jigs
- Improper personal management
- Not enough lighting
- Not enough air pressure and hot water or steam
- Poor maintenance
- Not enough skills in technical support
- Delay in raw materials supply
- Unexpected weather conditions
- Luck of R&D activities
- No data collection and analysis
- Daily changes in financial market

2.12 STEPS IN LONG-TERM PLANNING TO SOLVE PRODUCTION LINE PROBLEMS

Long-run accommodations to alter the production line and improve the products in a cost-effective manner as needed require the following steps:

1. Improving product design
2. Changing production line elements and handling facilities
3. Implementing 5S's in daily practices
4. Applying high standards and quality control
5. Implementing a modified version of the product
6. Training employees
7. Using new materials with high standards
8. Utilizing automation whenever it is possible

2.13 OBJECTIVES FOR LONG-TERM PLANNING TO IMPROVE THE PRODUCTION LINE

The key characteristics of short-run responsiveness and long accommodation are the aspects of flexibility and that they are instrumental in achieving the objectives of reduced labor, improved machine use, improved operational control, and reduced inventory. These objectives can help potential production line users in achieving the goals of lower product cost and higher quality. Below are the methods of achieving these objectives.

2.13.1 Reduced Labor

Labor is one of the major factors in cost calculation of the product. This cost can be reduced by the following:

- Maximize operator working times on machines.
- Utilize the devices, systems, and machines efficiently.
- Train labor to reduce the number of failures and rejects.
- Eliminate dependence upon highly skilled operators.
- Provide handling facilities to support unmanned operation.

2.13.2 Improved Machine Utilization

The proper use of machines, tools, jigs, devices, and systems will reduce the production time and achieve a perfect product. Therefore, to achieve this objective, the following can be adopted:

- Eliminate heavy and complicated machine setup.
- Use efficient energy consumption machines.
- Utilize automated features to replace manual intervention.
- Provide quick transfer devices to keep the machines in the production cycle.

2.13.3 Improved Operational Control

To achieve this objective the following can be adopted:

- Reducing the number of uncontrolled variables
- Providing tools to recognize and react quickly to deviations from the plan
- Reducing dependence upon human communication
- Implementing the lean manufacturing process whenever possible

2.13.4 Reduced Inventory

To achieve this objective the following can be adopted:

- Reducing lot sizes and inventory
- Improving turnover
- Providing the planning tools for just-in-time production
- Depending on reliable suppliers

2.13.5 Procedures

Each production line has many procedures to control the manufacturing steps of a product. Details and examples are given in Chapter 3. The procedures can be:

- Manufacturing procedures
- Test procedures
- Operation procedures
- Handling procedures
- Quality control procedures
- Chemical handling procedures
- Packaging procedures
- Shipping procedures
- Inventory procedures
- Internal and external auditing procedures
- Safety procedures
- Health procedures
- Special procedures

3 Manufacturing Procedures

3.1 TYPES OF MANUFACTURING PROCEDURES

Each production line has many procedures to control the manufacturing steps of a product. Manufacturing procedures can be listed as follows:

1. Manufacturing procedures
2. Test procedures
3. Operation procedures
4. Handling procedures
5. Quality control procedures
6. Chemical handling procedures
7. Packaging procedures
8. Shipping procedures
9. Inventory procedures
10. Internal and external auditing procedures
11. Safety procedures
12. Health procedures
13. Special procedures

3.2 SAMPLES OF MANUFACTURING PROCEDURES

Two samples of manufacturing procedures that are used in manufacturing an optic fiber device are given below.

3.2.1 Sample 1: Temperature Test (Manual Oven) for Hybrid Devices

1. **Purpose:** This procedure defines the method of temperature test for optical devices.
2. **Scope:** This procedure shall apply to optical devices that require performance temperature tests.
3. **Revision control:** When any part of this procedure requires amendment, the document shall be reissued in its entirety. Requests for change shall be addressed to the RR department.

Revision	Date	Reason	Originator
A	January 2015	First issue	John, Jack

4. **Definitions/abbreviations:**
   Temperature controller: Heat/cool manual controller.
5. **References:**
   ENG-VI-3456: Polarization Dependent Loss (PDL) Measurements for Optical Devices.
6. **Detailed description:**
   a. Switch on the temperature controller and set oven temperature to 23°C. Reference the PS meter output based on document ENG-VI-3456 on a regular basis as required, ensuring accurate reading.
   b. Put the hybrid device to be tested into the device holder inside the temperature controller.
   c. Check worksheet for device input and output port of insertion loss. Put output fiber to a fiber plug and put the fiber plug into the PS meter detector.
   d. Connect the connector of a pigtail to the PS meter output source. Use mechanical splicer to connect the pigtail to the input fiber of the device. Make sure the PDL reading displayed on the PS meter is lower than the higher of the PDL test station and PDL at cutback recorded on the worksheet.
   e. Set up another device on the oven by repeating steps a to d. Use a folded double-sided tape to separate the devices. Test no more than four devices at a time.
   f. Record the loss/PDL reading on the worksheet at 23°C for all devices.
   g. Set the oven temperature to the next temperature listed on the worksheet. Usually, the second temperature is the minimum temperature. Wait for 15 minutes. Record readings on the worksheet for all devices.

h. Set the oven temperature to the next temperature listed on the worksheet. Usually, the third temperature is the maximum temperature. Wait for 15 minutes. Record readings on the worksheet for all devices.
i. Set the oven temperature back to 23°C. Disconnect all the devices and place them back in the petri dishes.
j. Follow the instructions on the worksheet to calculate the difference (delta) loss/PDL at different temperatures. Based on the criteria listed on the worksheet, decide if the device is pass or fail.
k. Fill out proper information on both the traveler and worksheet. Complete the logbook for tested devices. Bring the tested devices to the quality assurance (QA) table.

3.2.2 Sample 2: AP and PC Connector Assembly

1. **Purpose:** This document describes the procedure for the manufacturing assembly of the AP and PC connectors.
2. **Scope:** This document shall apply to the manufacturing of the optical connector assembly of 250 and 900 microns on the fiber optic cable.
3. **Revision control:**

Revision	Date	Reason	Originator
A	February 2014	First release	John, Jack

4. **Definitions/abbreviations:**
Loss: The power difference between the initial power reading and final power reading through the device.
Cleaning swab: TX581 swab (P/N: P0999138).
Tweezers: A pair of tweezers (N: 333094) use for rotating and holding the tube and centerpieces, respectively.
Common laser: Laser used to align the signal port. Typical value: 1,550 nm.
5. **References:**
WPAP: Procedure for Applying Promoter.
WFS14: Fiber Stripping.
WCL19: Clean Lenses.
WTLJ: Tube to Lens Joint.
WS77: Workmanship Standard 77—Lens Polish Surface Quality.
WTPS: Tube Polish Surface Quality.
6. **Procedures:**
a. Materials setup:
i. Select the appropriate ring magnet (P/N: 8885) and a polished tube (P/N: WT22 or WT33) according to the product worksheet.
ii. Insert the magnet around the fibers. Make sure that the magnet orientation is the same as the one on the magnet used on the alignment jig. This can be verified by sticking the circular magnet to the one

on the jig. The arrow on the circular magnet should point in the direction of the light. It may be necessary to flip the magnet on the jig depending on the device.

iii. Inspect the tube surface as per L556. Clean the surface of the tube using ethanol and a cleaning swab.

iv. Select a tap or band pass centerpiece with a filter and an isolator corresponding to the specifications on the subassembly traveler.

v. Inspect the centerpiece for scratches, chipped corners, dirt, or stains. The polished surface should extend across the entire face of each lens. Clean or reject as per WTPS. Place rejected centerpieces back into their original centerpiece capsules and fill out the traveler accordingly.

vi. Clamp in place the AB or BC centerpiece on the alignment jig with the isolator located on the right-hand side, just over the magnet (see Figure 3.1). Make sure that the full length of the isolator is located on top of the magnet and that the epoxy joint between the two lenses is in the middle of the V-groove.

b. Tube alignment:

i. Splice the input fiber from the (double-bore) tube (port 1) to a SMF pigtail with an FC-PC connector (see Figure 3.2); obtain a loss of <0.3 dB. Plug in the connector to the common laser and measure the power at the output of the tube. Record the reading of the power meter on the product worksheet for the common wavelength in the "input tube" column.

ii. Clamp the fiber tube on the left alignment arm of the alignment jig. Make sure not to overtighten the clamp. The tube should be clamped at approximately 2 mm from its angled face.

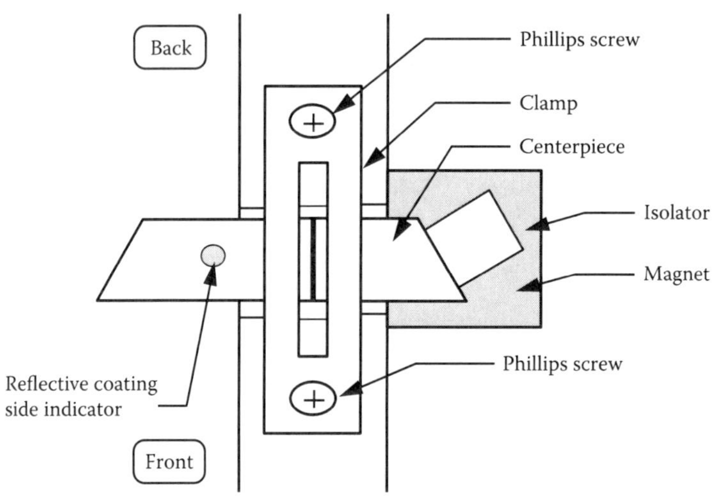

FIGURE 3.1 Layout of the centerpiece on the alignment jig.

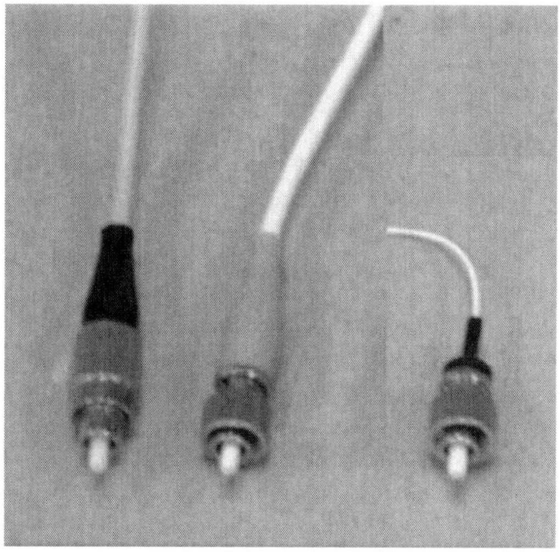

FIGURE 3.2 FC-PC connectors.

  iii. Align the tubes to the centerpiece with the three positioners on the alignment jig until the tubes are concentric (i.e., centered with each other) with the centerpiece. Make sure that the angled faces of the centerpiece and the tube match correctly (i.e., parallel). If the angles do not match, repeat step b.

c. Cleaning and visual inspection on the right side:
   i. Clean the polished tube and centerpiece surfaces with ethanol and a cleaning swab. Make sure the surfaces are void of stain or residue.
   ii. Clamp the fiber tube on the right alignment arm of the alignment jig. Make sure not to overtighten. Also, make sure that only the tube is clamped and not the centerpiece.
   iii. Align the centerpiece to the isolator with the three positioners. Make sure that the angles of the centerpiece and the isolator match correctly.
   iv. Clean the isolator and centerpiece surfaces with ethanol and a clean swab. Make sure the surfaces are void of stain or residue.
   v. Close the gap between the tube on the left side and the lens until the tube touches the centerpiece. Record the reading of the (micrometer) positioner. This is the value when the gap between the tube and the centerpiece is zero.

d. Alignment on the left side:
   i. Connect the fiber of port 2 to the power meter.
   ii. Press the "channel" button to switch the meter reading to the left side.
   iii. Adjust the position of the (double) fiber tube with the positioner on the jig to find the minimum loss. When the loss is approximately 2 dB or less of the expected value optimize (see Figure 3.3) by pressing the "optimize" button on the auto aligner.

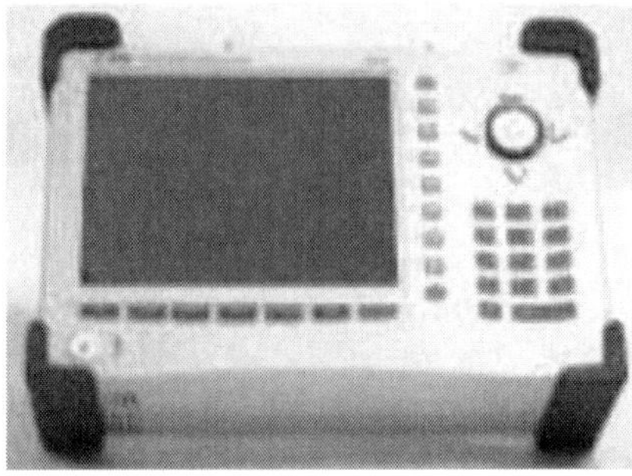

FIGURE 3.3 Auto aligner.

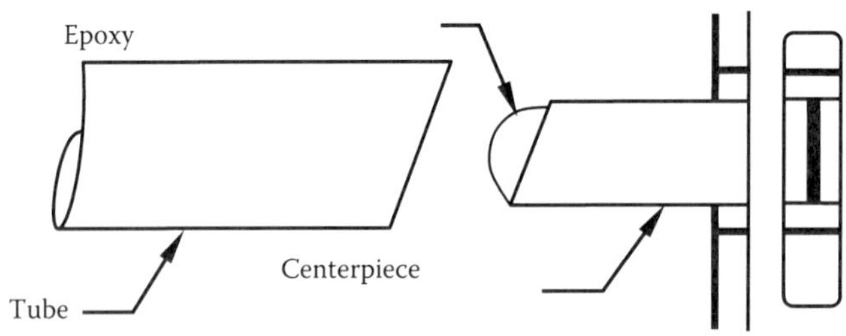

FIGURE 3.4 Epoxy on the centerpiece and tube.

iv. Push back the arm and clean the lens using a cleaning swab and denatured ethanol. Apply adhesive promoter solution to the lens and the tube using a swab according to 669812.
v. Apply a small drop of LJ699 epoxy to the center of the centerpiece surface (see Figure 3.4).
vi. Bring the tube close to the centerpiece until the epoxy forms a bridge between the tube and the centerpiece. Optimize with the positioners on the alignment jig.
vii. Using the "optimize" button on the auto aligner (see Figure 3.3), monitor the glue gap to get the lowest loss on the power meter.
viii. Record the reading of the power meter on the product worksheet in the "glue" column after the epoxy was introduced. Make sure the gap between the tube and the centerpiece is between 10 and 30 μm. If the gap is not within that range, replace the centerpiece beside the isolator assembly and repeat steps c and d.

e. Insertion loss alignment on the right:
   i. Close the gap between the isolator and the lens on the right-hand side until the lens touches the isolator and record the reading on the (micrometer) positioner. This is the value when the gap between the isolator and the lens is zero. *Be careful not to force the tube into the centerpiece. If it is forced, it may damage the device.*

  ii. Connect the output fiber of the tube on the right-hand side of the jig to the power meter. Connect the input fiber (port 1) on the left-hand side to the common laser. **Note:** For AB, the device is upside down (such that the isolator is on the right) on the jig, so the input fiber is on the right.
  iii. Adjust the position of the fiber tube on the right-hand side with the positioner on the jig to find the minimum loss. When the loss is approximately 2 dB or less, optimize with the auto aligner (see Figure 3.3).
  iv. Apply adhesive promoter solvent (A359815) to the lens and the isolator using a clean swab according to LIO12.
  v. Apply a small drop of LJ689 epoxy to the center of the centerpiece surface. Bring the lens close to the isolator until the epoxy forms a bridge between the lens and the isolator. Optimize with the positioners on the alignment jig.
  vi. Monitor and adjust the fine adjustment positioners on the auto aligner (see Figure 3.3) and the glue gap on the right-hand side to get the lowest loss on the power meter. Make sure the gap between the isolator and the tube is between 10 and 35 μm. If the gap is not within that range, replace the centerpiece with a (double-bore) tube assembly with another and repeat steps d and e.
  vii. If the specification for insertion loss cannot be met, replace the centerpiece with the (double) fiber tube with another and repeat steps d and e. Fill out the form to accompany the rejected part.
  viii. Record the reading of the power meter on the product worksheet in the "glue" column, insertion loss as well as the glue gap.
  ix. Record the reading on the power meter for the common wavelength (tap loss).

f. Right cure:
  i. Connect the fiber of port 1 to the common laser and the fiber of port 2 to the power meter (if not already connected).
  ii. Turn on the auto aligner and press the "heater" button.
  iii. Press "optimize" using the auto aligner.
  iv. When the "heat on" light goes off, the joint is cured. The "remote on" light should also have gone off.
  v. Release the top clamp.
  vi. Lower the clamp of the alignment arm on the right.

g. Left cure:
  i. Place the curing arm over the glue joint to be cured on the left-hand side (tube to centerpiece).
  ii. Press the "channel" button to switch the meter reading to the left, i.e., channel 1.
  iii. Connect the fiber of port 1 to the common laser and the fiber of port 2 to the power meter if it is not already connected.
  iv. Turn on the auto aligner and press the "heater" button.

   v. When the "heat on" light goes off, the joint is cured. The "remote on" light should also have gone off.
   vi. Release the top clamp.

h. Final loss measurements:
   i. Record the reading of the power meter for the insertion loss.
   ii. Cut the input fiber 15 cm after the mechanical splice and measure the loss of the input fiber. Record the value on the subassembly traveler (input fiber). Subtract the output fiber loss from the input fiber loss, and report this value as the cutback reflected loss in the "loss" column.
   iii. If the difference between the cutback reflected loss and the reflected loss measured at assembly after curing is greater than 0.1 dB, perform a second cutback. Repeat the cutback process until you get two cutbacks within 0.05 dB.
   iv. Put the subassembly and the filter plot if available in a petri dish along with the traveler.
   v. Color the fibers according to the specifications on the product worksheet.
   vi. Put the device in a petri dish with the corresponding worksheet.

APPENDIX 3A: PDL TEST PROCEDURE

1. Perform PDL tests for all of the ports indicated on the PDL section of the device worksheet.
2. Strip the output port and insert it into the fiber holder. Cleave the fiber to approximately 1 mm beyond the fiber holder and insert it into the fiber plug on the PDL meter. (Note: The device may already be connectorized, in which case it only needs to be plugged into the meter.)
3. Make sure the magnet is on the isolator.
4. Splice and cleave the input fiber and insert it into the splice box.
5. Press the PDL button on the meter and the average loss, followed by the PDL, will be displayed. Cleave and resplice both fibers until good agreement in loss is reached between successive cleaves. The loss should be <0.3 dBm than loss indicated on the worksheet.
6. Pass or fail the device according to the specifications on the worksheet, and fill out the traveler and worksheet appropriately. Include initials, date, and PS meter number.
7. Place the passed device in a glass petri dish and send for overnight baking.

APPENDIX 3B: TROUBLESHOOTING AND POSSIBLE FIXES

Note: These are only possible reasons and solutions to problems that may be encountered. If difficulty arises during alignment, your supervisor/lead hand should be notified of the problem.

Unable to optimize before applying epoxy (loss > 2 dB):

1. Recleave output fiber.
2. Check alignment.
3. Clean tube and centerpiece surfaces.

Unable to optimize after applying epoxy (loss > 0.4 dB):

1. Recleave output fiber.
2. Check alignment.
3. Possible contamination (dirt) or air bubbles in epoxy.
4. Remove epoxy and check tube to centerpiece alignment for angles.
5. Check for change in input tube reading.

Good loss with large gaps:

1. Lens length on centerpiece may be incorrect.

The loss goes up during cure:

1. Device not optimized properly to lowest possible value.
2. Tube too close to a centerpiece (gap too small).
3. Tube incorrectly aligned for angle.
4. Air bubble expanding with heat.

The loss goes up during cooling:

1. Device not fully cured.
2. Misalignment (if problem persists have jig realigned).
3. Device not allowed to cool sufficiently.

The loss goes up after releasing tube or centerpiece clamp:

1. Clamp is too tight.
2. Device not fully cured.

4 Clean Rooms

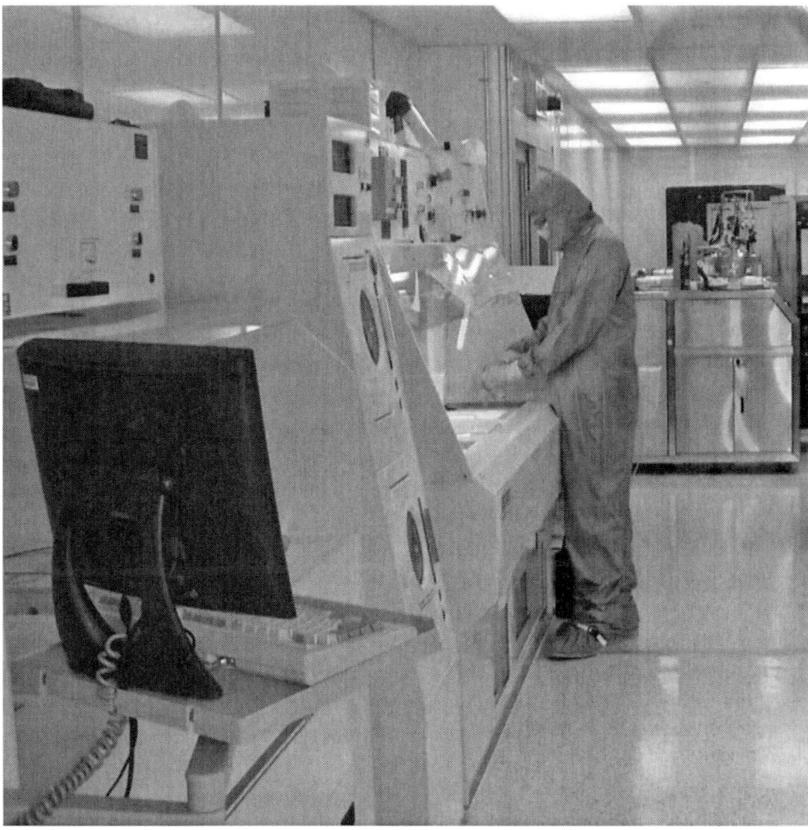

4.1 INTRODUCTION

With the advent of microprocessors and photonics devices, a need for specialized manufacturing arose. The reason processor speeds have been constantly increasing is because the size of the physical components has been decreasing. Since the size of processor conductors has decreased into the nanometer range, an ultra-clean environment to produce these chips has become necessary. Fabricating microprocessors was the original purpose for clean rooms.

A clean room is a very clean environment that has an extremely low level of pollutants such as dust, airborne microbes, aerosols, humidity, and chemical vapors. These types of contaminants are controlled by various means. Bunny suits, extensive filtration systems, and a strict policy of what can and cannot be brought into the clean room keep the facility dust- and particle-free. Temperature and humidity are

FIGURE 4.1 A clean room in a production line.

also closely monitored to prevent other various types of damage to the product during manufacturing.

When one thinks of how large a particle of dust is, extremely small might be a description that comes to mind. In manufacturing, it is quite the opposite. Dust can be the size of a giant boulder compared to the even smaller product that is being manufactured. Imagine the destruction if Godzilla decided to take a walk down the streets of downtown Ottawa. This is the case if a dust particle landed on a circuit board or a piece of nanotechnology.

These dust particles are controlled in a very precise manner, as shown in Figure 4.1. First, the human contaminants must be stopped. Removing humans all together from the fabrication process would be ideal, but this is impossible, as automation has not reached that level. Humans must be used. The human body releases on average about 25,000 particles per minute. On top of that, regular street clothing releases up to 10 million particles per minute during normal walking conditions. There is one more area that humans expel particles from. Aerosols can be released from breathing or coughing (a cough produces roughly 600,000 droplets into the atmosphere). The main process to inhibit the release all of these contaminants into the production environment is to get the humans involved in the production line to wear bunny suits.

4.2 BUNNY SUIT

Protective clothing must be worn by someone working in a clean room to prevent dust, human skin, and hair particles from entering the room's atmosphere. This will protect sensitive equipment from contamination.

Clean rooms are important features in the production of silicon chips, hard disk drives, and other technologies, such as satellites, foods, chemicals, and in pharmaceutical production. The air in a clean room is repeatedly filtered to remove dust particles and other impurities that can damage the production of highly sensitive technologies. The measure of the air quality of a clean room is described in Federal Standard 209. Clean rooms are rated as Class 10,000, where there exists no more than 10,000 particles larger than 0.5 microns in any given cubic foot of air; Class 1,000, where there exists no more than 1,000 particles; and Class 100, where there exist no more than 100 particles. Hard disk drive fabrication requires a Class 100 clean room.

Special protective clothing, sometimes called a bunny suit, does not give off lint particles and prevents human skin and hair particles from entering the room's atmosphere. Figure 4.2 shows a bunny suit can comprise many, if not all, of the following parts:

- Surgical mask
- Lint-free head cover
- Shoe covers
- Booties
- Gowning gloves
- Bunny suit overcoat and pants
- Latex gloves
- Belt
- Filter unit
- Battery pack
- Helmet
- Safety glasses
- Scope shield

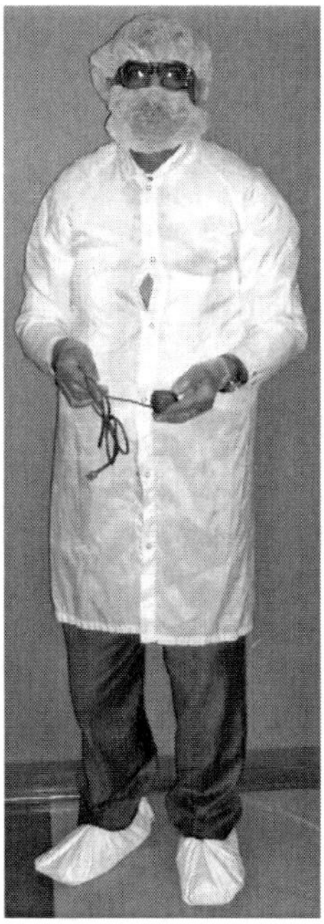

FIGURE 4.2 A bunny suit.

These articles all must be donned in a very specific order to maintain the quality of the clean room. The fabric used in all of those aforementioned places is another concern to the cleanliness of the room. Conventional natural fabrics are not strong enough to be woven tightly to achieve no penetration of contaminants. They also have natural contaminants that get stuck in the pores of the material. Therefore, natural fabrics cannot be used. Instead, synthetics were introduced into the clean room. These are much stronger, which results in a tighter weave. This leaves fewer gaps for particles to get transferred through. Synthetic fibers do not, however, transmit electricity. This causes a large problem because if enough static electricity builds up, the smallest jolt of electricity traveling into ground can seriously damage the equipment. This problem is fixed by weaving microstrands of aluminum into the bunny suit. Another way to reduce airborne particles is to ensure the maximum filtration of the air. Many clean rooms have the entire volume of air circulated every few minutes. The air passes through many HEPA filters, which are capable of trapping large and even extremely small particles. Also, the clean room is slightly pressurized so that if there are any leaks, the air will travel from inside to out, preventing the inward flow of outside contaminants.

4.3 CLEAN ROOM STANDARDS

Despite all of these protective measures, all clean rooms are not created equal. There are different standards that certain clean rooms must adhere to depending on the sensitivity of the product. Tables 4.1 and 4.2 present two different standards and their accompanying levels of cleanliness. They show the ISO 14644-1 and US Federal Standard 209E ISO 14644-1 for clean rooms.

The most common scale is the ISO 14644-1 standard, since US Federal Standard 209E was discontinued on November 29, 2001.

Clean rooms are widely used in many industries, including space exploration, computer manufacturing, nanotechnology, pharmaceutical industries, and especially

TABLE 4.1
ISO 14644-1: Clean Room Standards

	Particles/m³					
Class	**0.1 μm**	**0.2 μm**	**0.3 μm**	**0.5 μm**	**1 μm**	**5 μm**
ISO 1	10	2				
ISO 2	100	24	10	4		
ISO 3	1,000	237	102	35	8	
ISO 4	10,000	2,370	1,020	352	83	
ISO 5	100,000	23,700	10,200	3,520	832	29
ISO 6	1,000,000	237,000	102,000	35,200	8,320	293
ISO 7				352,000	83,200	2,930
ISO 8				3,520,000	832,000	29,300
ISO 9				35,200,000	8,320,000	293,000

TABLE 4.2
US Federal Standard 209E ISO 14644-1: Clean Room Standards

	Particles/ft³				
Class	**0.1 μm**	**0.2 μm**	**0.3 μm**	**0.5 μm**	**5 μm**
1	35	7	3	1	
10	350	75	30	10	
100		750	300	100	
1,000				1,000	7
10,000				10,000	70
100,000				100,000	700

optical manufacturing. All of these fields are essential to maintain the quality of life that we are used to, and without clean rooms, none of this would be possible. Using clean rooms in manufacturing sophisticated devices will add cost to a product. It is impossible to produce these devices without using clean rooms. Therefore, the cost associated with using clean rooms is one of the final cost elements.

5 High Manufacturing Technology

5.1 INTRODUCTION

This chapter presents some examples of high manufacturing techniques that are used in manufacturing very sophisticated and highly accurate dimension devices. These devices are used in building telecommunication, medical, military, testing, and scanning systems.

The interconnection of optical components is a vital part of an optical system, affecting performance. With a light source, optical components, and fiber optic cables, a fiber optic device can be built. A fiber optic system can be created when more optical fiber devices are aggregated together.

This chapter also presents fiber optic cable types and manufacturing processes. It is very important to learn fiber optic cable installation techniques. Appropriate fiber optic installation avoids cable losses. Fiber cables can be installed on land and undersea. Fiber optic cable applications are presented throughout the book.

Interconnection between two fiber optic cables is achieved by either connectors or splices, which link the ends of the fiber cables optically and mechanically. Connectors are devices used to connect a fiber optic cable to an optical fiber device, such as a detector, optical amplifier, or optical light power meter, or link to another fiber cable in the same mode. They are designed to be easily connected and disconnected reliably with constant losses. The connectors create an intimate contact between the mated halves to minimize the power loss across the junction. They are appropriate for indoor applications. Splices are used to connect one fiber optic cable to another permanently. Splices are suitable for outdoor and indoor applications. Some types of splices are used to temporarily connect, for quick testing purposes, during manufacturing processes.

5.2 FIBER OPTIC CABLES

Fiber optic cables transmit data through very small core diameters at the speed of light. They are significantly different from copper cables. Fiber optic cables offer high bandwidths, low losses, and allow high data transmission rates over long distances. Light propagates throughout the fiber cables by the principle of total internal reflection. The idea of a waveguide is to guide the light in an optical system. Therefore, fiber optic cable is also called waveguide, like any waveguide design.

There are three common types of fiber optic cables: single-mode, multimode, and graded-index cables. Each has its advantages and disadvantages. The advantages and disadvantages are compared to modes, transmission rates, installation, and costs. There also are several different designs of fiber optic cables, which are made for different applications. New fiber optic cables with different core and cladding designs are emerging in industry. The new designs can carry more modes and higher transmission data rates than the common ones. Fiber optic cables are used mostly in communication and data collection systems. Other applications, such as medical, military, scanning, imaging, sensing, optical fiber devices, and fiber optic lighting, are very well established.

This chapter presents the fiber optic management and fabrication processes used in manufacturing fiber cables. The processes produce a flexible glass strand of unbelievably small diameter, thinner than human hair.

5.3 FIBER OPTIC CABLE CONSTRUCTION

Fiber optic cable is a filament of transparent material used to transmit light, as shown in Figure 5.1. Virtually all fiber optic cables share the same fundamental structure.

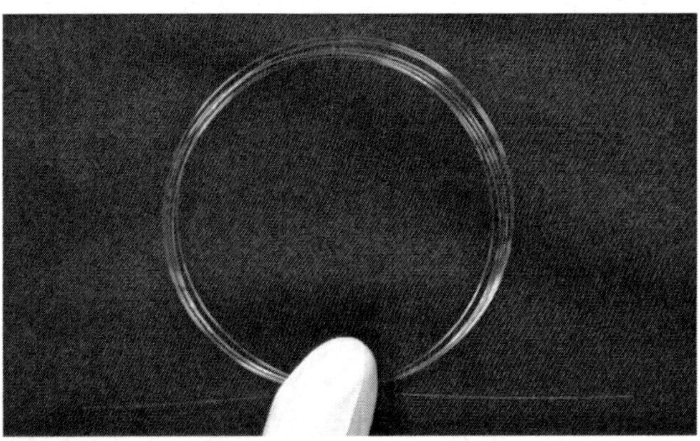

FIGURE 5.1 A fiber optic cable.

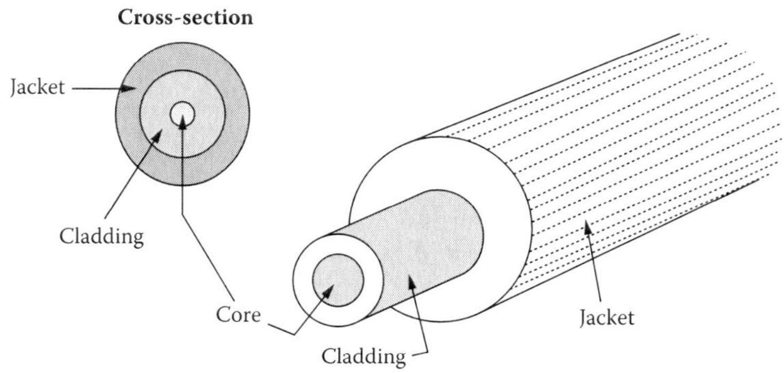

FIGURE 5.2 Schematic view and cross section of a fiber optic cable.

The center of the fiber optic cable is referred to as the core. The core has a higher refractive index than the cladding, which surrounds the core. The contact surface between the core and the cladding creates an interface surface that guides the light. The difference between the refractive index of the core and cladding causes a mirror-like interface surface, which guides light along the core. Light bounces through the core from one end to the other by the principle of total internal reflection, as explained by the laws of light. The cladding is then covered with a protective plastic or PVC jacket. The diameters of the core, cladding, and jacket can vary widely. For example, a single fiber optic cable has core, cladding, and jacket diameters of 9, 125, and 250 μm, respectively.

Figure 5.2 shows the structure of a typical fiber optic cable. The cores of most fiber optic cables are made from pure glass. The claddings are made from less pure glass. Glass fiber optic cable has the lowest attenuation over long distances and comes at the highest cost. A pure glass fiber optic cable has a glass core and a glass cladding. Fiber optic cable cores and claddings may be made from plastic; plastic is not as clear as glass but is more flexible and easier to handle. Compared with other fiber cables, plastic fiber cables are limited in power loss and bandwidth. However, they are more affordable, easy to use, and attractive in applications where high bandwidth or low loss is not a concern. A few glass fiber cable cores are clad with plastic. Their performance, though not as good as all glass fiber cables, is quite respectable. More details on plastic cables are presented later in this chapter.

The jacket is made from polymer (PVC, plastic, etc.) to protect the core and the cladding from mechanical damage. The jacket has several major attributes, such as bending ability, abrasion resistance, static fatigue protection, toughness, moisture resistance, and ability to be stripped. Fiber optic cable jackets are made in different colors for color-coding identification. Some optical fibers are coated with a copper-based alloy that allows operation up to 700 and 500°C for short and long periods, respectively.

Table 5.1 shows the common diameters of the core, cladding, and jacket of four commonly used fiber optic cables in manufacturing optical fiber devices. Different types and cross sections are also available for building advanced fiber optic devices. These types will be presented in the coming sections of this chapter.

TABLE 5.1
Common Diameters of Fiber Optic Cables

Standard Diameter of		
Core (μm)	Cladding (μm)	Jacket (μm)
8	125	250
50	125	250
62.5	125	250
100	140	250

5.4 PLASTIC FIBER CABLES

Plastic optical fiber (POF) has the highest attenuation over short distances, but comes at the lowest cost. A plastic fiber optic cable has a plastic core and plastic cladding. This fiber optic cable is quite thick. Typical diameters of core/cladding are 480/500, 735/750, and 980/1,000 microns. The core generally consists of polymethylmethacrylate (PMMA) coated with a fluoropolymer. Plastic fiber optic cables are used in small optical devices, lighting applications, automobiles, music systems, and other electronic systems. They are also used in communication systems where high bandwidth or low loss is not a concern. The increased interest in plastic optic fiber is due to two reasons. First, the higher attenuation relative to glass may not be a serious obstacle with the short cable runs often required in premise networks. Second, the cost advantage sparks interest when network architects are faced with budget decisions. Plastic fiber optic cable does have a problem with flammability. Because of this, it may not be appropriate for certain environments and care has to be given when it is run through a plenum. Otherwise, plastic fiber is considered extremely rugged with a tight bend radius and the ability to withstand mechanical stress.

Plastic clad silica (PCS) fiber optic cable has an attenuation that lies between glass and plastic and a cost that lies between their costs as well. PCS fiber optic cable has a glass core that is often vitreous silica, while the cladding is plastic, usually a silicone elastomer with a lower refractive index. In 1984 the International Electrotechnical Commission (IEC) standardized PCS fiber optic cable to have the following dimensions: core, 200 microns; silicone elastomer cladding, 380 microns; jacket, 600 microns.

Plastic fiber cables are fabricated using the same principles as glass fiber cables. A core with a higher index of refraction is surrounded by a cladding with a lower index of refraction. The cladding is coated with colored jacket for color-coding purposes. Glass and plastic cables have similar color coding. POF cables are available in single- and multimode step index and graded index.

However, there have been some improvements in plastic fiber optic cable in recent years due to the developments in the polymer industry. Plastic fiber cables will replace the glass fiber cables because of many advantages, such as ease in connection using epoxy, lower price, durability, lower weight, and smaller bending radii.

5.5 LIGHT PROPAGATION IN FIBER OPTIC CABLES

Figure 5.3 shows the principle of light propagation through a fiber optic cable. Consider light injected in an optical material, which is surrounded by another optical material with a different index of refraction. The first and second optical materials have a higher and lower index of refraction, respectively. Figure 5.3(a) shows light incident at an angle θ to the normal. The light refracts by an angle β from the normal. If the angle of incidence is increased, the angle of refraction will increase also, as shown in Figure 5.3(b). In Figure 5.3(c), light incident at an angle θ equal to the critical angle θ_c gives an angle of refraction β of 90°. The refracted light lies on the interface of the first-second optical materials. Snell's law is applicable in the above cases. In Figure 5.3(d), when light lands incident at an angle θ greater than the critical angle, the light will reflect by the total internal reflection phenomena. In this case, the angle of incidence equals the angle of reflection, which is defined by the first law of light.

Assume the first optical material is very long and has a circular cross section with a very small diameter like the core, and at the same time it is surrounded by another optical material like the cladding in a fiber optic cable, as shown in Figure 5.3(e). Now, light injected into a fiber optic cable and striking the core-cladding interface at a greater angle than the critical angle reflects back into the core. Since the angles of incidence and reflection are equal, the reflected light will again be reflected. The light will continue to bounce through the length of the fiber cable. Cases (a), (b), and (c) in Figure 5.3 show that the light passes into the cladding. The cladding is usually inefficient as a light carrier compared to the core of the fiber cable. Light in the cladding becomes part of the losses that usually occur in any fiber optic cable. Some of the common losses will be presented in the coming sections. Therefore, light propagation in a fiber optic cable is governed by the following:

- The wavelength of light
- The angle of incidence of the light at the input of the fiber cable
- The indices of refraction of the core and cladding
- The composition of the core and cladding
- The length of the cable
- The bending radius of the cable
- The sizes of the core and cladding
- The design of the core and cladding
- The transmission modes
- The temperature and environment conditions of the fiber cable
- The strength and flexibility of the fiber cable
- The installation methods

5.6 TYPES OF FIBER OPTIC CABLES

As explained above, there are three common types of fiber optic cables, listed below. The suitability of each type for particular applications depends on the fiber optic cable characteristics.

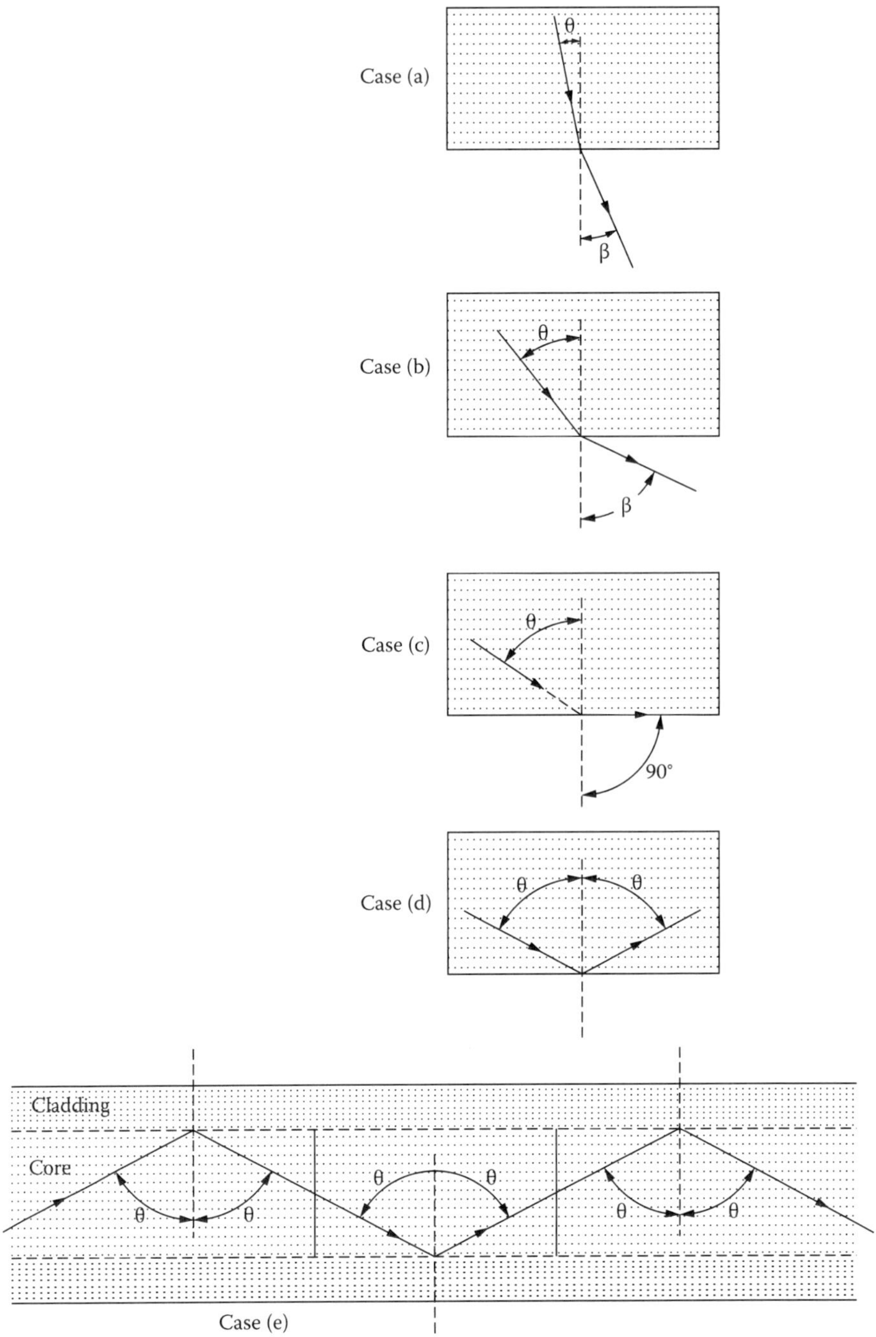

FIGURE 5.3 Principle of total internal reflection in a fiber optic cable.

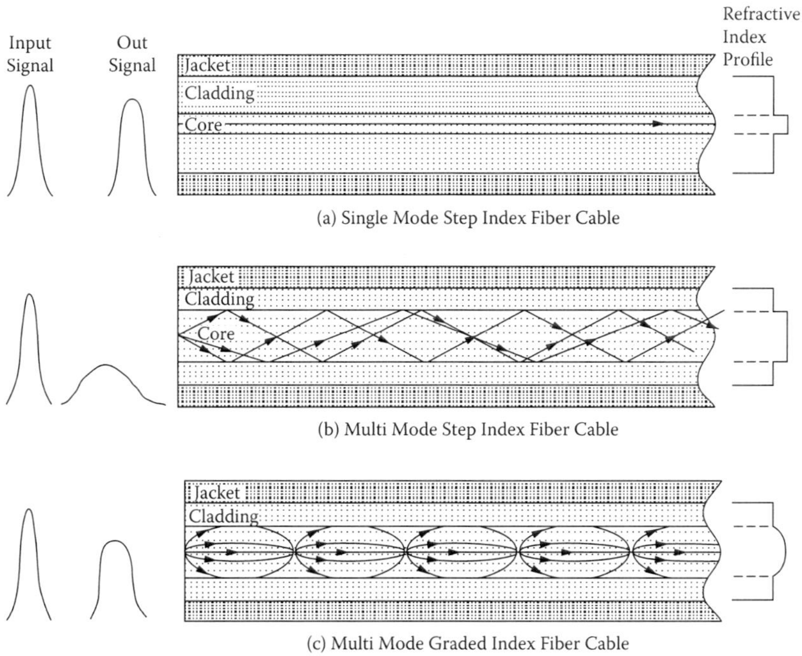

FIGURE 5.4 Types of fiber optic cables.

5.6.1 Single-Mode Step-Index Fiber Cable

This fiber optic cable type, sometimes called single-mode fiber cable, is shown in Figure 5.4(a). The single- and multimode step-index fiber cables are the simplest types. Single-mode fiber cables have extremely small core diameters, from 5 to 9.5 μm. The core is surrounded by a standard cladding diameter of 125 μm. The jacket is applied on the cladding to provide mechanical protection, as shown in Figure 5.4. Jackets are made of one type of polymer in different colors for color-coding purposes. Single-mode fibers have a potential to carry signals for long distances with low loss. They are mainly used for communication systems. The number of modes that propagate in a single-mode fiber depends on the wavelength of light carried. As the wavelength is increased, the fiber carries fewer and fewer modes until only one remains. Single-mode operation begins when the wavelength approaches the core diameter. At 1,310 nm, for example, the fiber cable permits only one mode. It operates as a single-mode fiber cable.

5.6.2 Multimode Step-Index Fiber Cable (Multimode Fiber Cable)

This fiber type, sometimes called multimode fiber cable, is shown in Figure 5.4(b). Multimode fiber cables have bigger diameters than the single-mode ones. They have a core diameter from 100 to 970 μm. They are available as glass fibers (a glass core and glass cladding), plastic clad silica (a glass core and plastic cladding), and plastic fibers (a plastic core and cladding). They are the widest ranging, although not the most efficient, in long distances and experience higher losses than the single-mode fiber cables. Multimode fiber cables have the potential to carry signals for moderate and long distances with low loss (when optical amplifiers are used to boost the signals to the required power).

Since light rays bounded through a fiber cable reflect at different angles for different ray paths, the path lengths of different modes are different. Thus, different rays take a shorter or longer time to travel the length of the fiber cable. The ray that goes straight down the center of the core without reflecting arrives at the other end faster. Other rays take slightly longer and thus arrive later. Therefore, light rays entering the fiber at the same time exit the other end at different times. The light has spread out in time. This spreading of light rays due to different modes is called modal dispersion. Dispersion describes the spreading of light rays by various mechanisms. Modal dispersion is that type of dispersion that results from the varying modal path lengths in the fiber cable.

5.6.3 Multimode Graded-Index (GRIN) Fiber Cable

This fiber type is sometimes called graded-index fiber cables and shortened to GRIN fiber cable, as shown in Figure 5.4(c). Graded-index and multimode fiber cables have similar diameters. Common graded-index fibers have core diameters of 50, 62.5, or 85 μm and a cladding diameter of 125 μm. The core consists of numerous concentric layers of glass, somewhat like the annular rings of a tree or a piece of onion. Each successive layer outward from the central axis of the core has a lower index of refraction until reaching the inner diameter of the cladding. Light travels faster in an optical material with a lower index of refraction. So the further the light is from the center axis, the greater is its speed. Each layer of the core refracts the light according to Snell's law. Instead of being sharply reflected as it is in a step-index fiber, the light is now bent or continually refracted in an almost sinusoidal pattern. Those light rays that follow the longest path by traveling near the outside of the core have a faster average velocity. The light ray traveling near the center of the core has the slowest average velocity. As a result, all rays tend to reach the end of the fiber at the same time. Thus, one way to reduce modal dispersion is to use graded-index fibers. This type of fiber optic cable is popular in applications requiring a wide range of wavelengths, especially in telecommunication systems, scanning, imaging, and data processing.

Fiber cables are designed for a specific wavelength, called the cutoff wavelength, above which the fiber carries only one mode. A fiber designed for single-mode operation at 1,310 nm has a cutoff wavelength of around 1,200 nm.

Although optical power is confined to the core in a multimode fiber, it is not so confined in a single-mode fiber. *Mode field diameter* is the term used to define this diameter of optical power. It is usually more important to know the mode field diameter than the core diameter.

5.7 POLARIZATION-MAINTAINING FIBER CABLES

Polarization-maintaining (PM) fibers are constructed by placing specially designed asymmetries into the core. PM fibers guide only one possible mode of propagation.

They also maintain the electromagnetic field vector direction. This type of single-mode fiber is used in building optical fiber devices that work with polarized light, such as polarization beamsplitters, couplers, modulators, and interferometric sensors. PM fibers are desirable since most lasers emit highly polarized light and the polarized properties are highly desirable in many measurement applications. There are four common designs for PM fibers shown in Figure 5.5. Figure 5.5(a) shows the PANDA fiber. The PANDA fiber employs a stress technique to stress the core of the fiber to create two propagation paths within the fiber core. Two stress rods are placed within the cladding in the same plane on opposite sides of the fiber core. Linearly polarized light aligned to either the slow or fast axis of the fiber will remain linearly polarized as it propagates through the fiber.

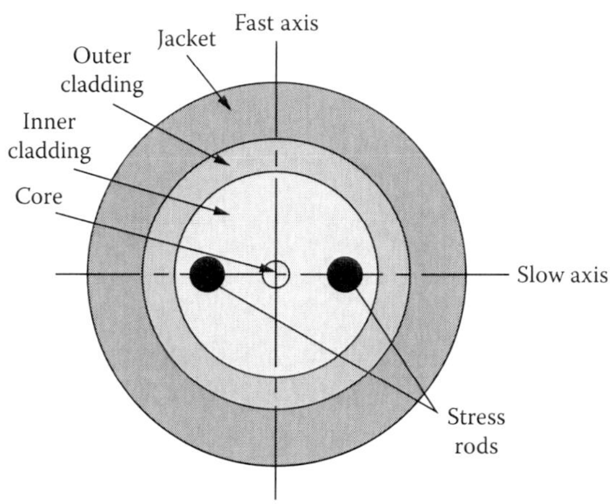

(a) Panda Fiber

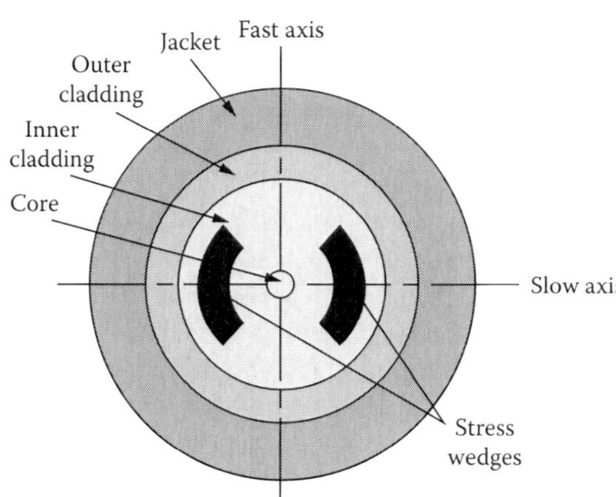

(b) Bow-tie-Fiber

FIGURE 5.5 Polarization-maintaining fiber cables. *(Continued)*

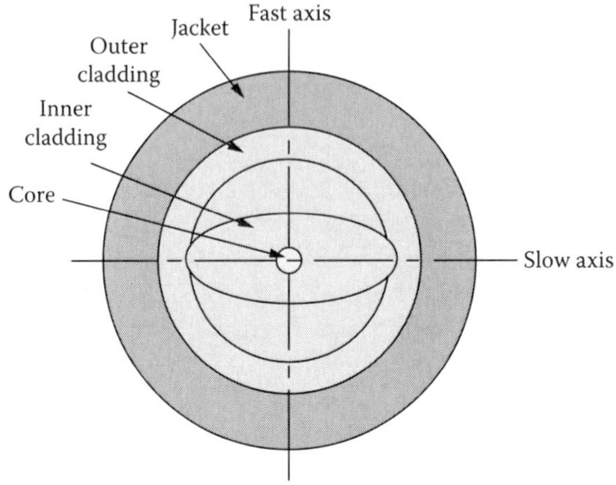

(c) Elliptical core fiber

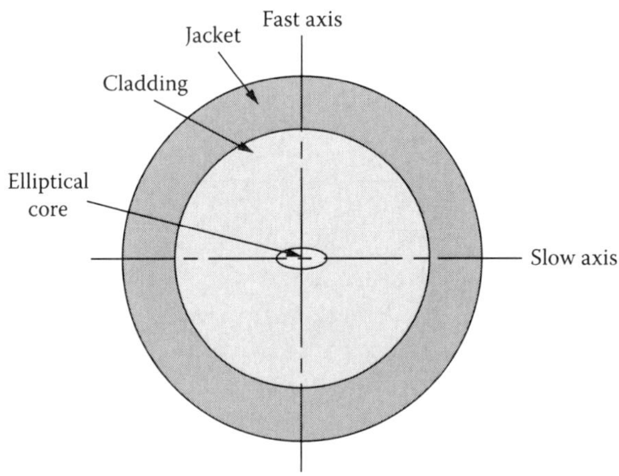

(d) Elliptical core

FIGURE 5.5 (Continued) Polarization-maintaining fiber cables.

Figure 5.5(b) shows a cross section of the bow-tie fiber. A pair of wedges on opposite sides of the core generate the optimum stress distribution within the fiber. This patented design gives the best in both performance and handling with minimum stress breakout when cleaved, connectorized, or polished.

Figure 5.5(c) shows a circular core surrounded with an elliptical boron-doped cladding. Figure 5.5(d) shows a PM fiber with a very high level of doping in an elliptical core, which causes waves polarized along the major and minor axes of the ellipse to have different effective indices of refraction. Thus, the fibers contain non-symmetrical stress production parts.

5.8 SPECIALTY FIBER CABLES

Previous sections presented common fiber cables that are used to guide light over relatively long distances in communication systems, imaging, scanning, and medical applications.

Fiber optic cable technology and applications have experienced a diversity of technological advancements that make optical fibers able to fulfill every possible application. Other types of fibers are optimized for a variety of specific applications. These types are also fabricated for research and development.

Chapter 6 presents the advanced fiber optic cables used in many applications requiring a low loss, low chromatic dispersion, and high transmission rate over long distances.

5.9 FIBER CABLE FABRICATION TECHNIQUES

There are a variety of fabrication techniques used to produce fiber optic cables. Two common methods used to manufacture optical glass fiber cables are presented in this section. The first method is directly drawing the fiber from two molten glass preforms, which are placed in two concentric crucibles; this is called the double crucible method. The second method is to form a glass rod called preform. The second method has several processes for producing preforms to be used in the fiber drawing method. The manufacturing of fiber cable requires sophisticated and highly accurate techniques. The biggest challenge facing fiber manufacturing is the purification of the materials used in the construction of the core and cladding. The value of index of refraction must be very precise. More precision is needed, especially when manufacturing graded-index fibers. During the glass fiber optic cable fabrication process, impurities are intentionally added to the pure glass to obtain the desired indices of refraction needed to guide light. Germanium or phosphorous is added to increase the index of refraction. Boron or fluorine is added to decrease the index of refraction. Unfortunately, adding these residual impurities may increase the attenuation by either scattering or absorbing light. The fabrication methods need to be extremely precise in fiber dimensions and tolerances. The diameter of the core must be controlled to high precision, and the core must be located at the center of the cladding. These same issues are also present when applying the cladding onto the core. In the final step, the cladding is coated with a polymer jacket layer for mechanical and environmental protection.

5.10 DOUBLE CRUCIBLE METHOD

The double crucible method is illustrated in Figure 5.6. A pair of platinum crucibles sit one inside the other. Molten core glass is placed in the inner crucible, and molten cladding glass is fed in the outer crucible. The crucibles are kept at a high temperature, typically between 1,850 and 2,000°C. Using a precision-feed mechanism, the two glasses come together at the base of the outer crucible, forming a core-cladding fiber. The fiber is drawn out from the crucibles. Then the fiber

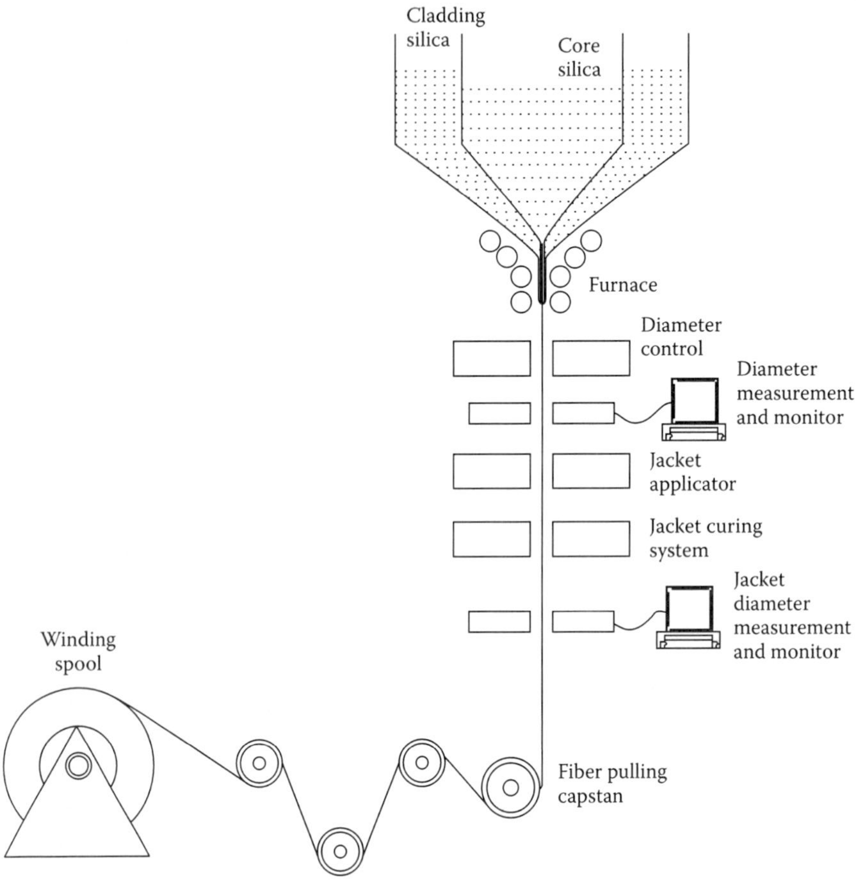

FIGURE 5.6 Double crucible fabrication method.

passes through high-precision diameter measurement and control equipment and is monitored by imaging or x-rays. The control equipment detects any nonhomogeneity and bubbles in the drawn fiber. The fiber is then covered by a colored layer of jacket. Again, the jacket layer goes through diameter control and monitoring equipment. The end of the fiber cable is attached to a rotating spool, which turns steadily. The fiber is then tested for attenuation (dB per kilometer), dispersion, and any other requirements specified by the customer or industry. Fiber cable lengths can be a few hundred kilometers, typically required by industry. Fiber cables are available in different types and working wavelengths.

The rod tube is one of the simplest methods of fiber fabrication. In the rod tube procedure, a rod of core glass is placed inside of a tube of cladding glass. This arrangement forms a preform required for the drawing process of the fiber. A preform is shown in Figure 5.7.

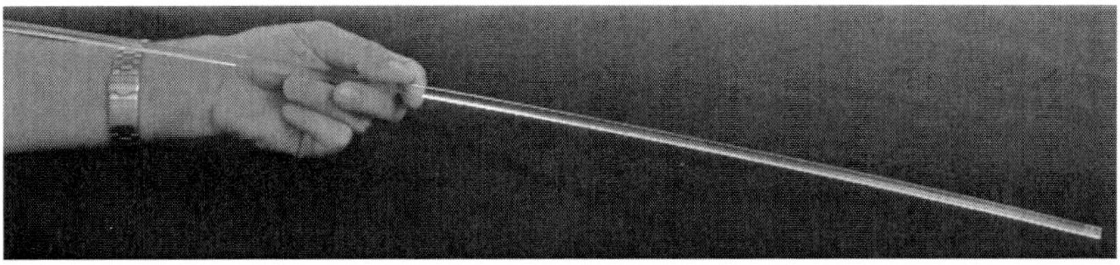

FIGURE 5.7 A preform.

5.11 CHEMICAL VAPOR DEPOSITION (CVD) PROCESSES

The preforms used in the fiber drawing method are fabricated using chemical vapor deposition (CVD) processes. The fabrication steps of the CVD method are similar to those explained in the double crucible method. A preform is needed for the fiber drawing process. The method for preparing a preform is called the rod tube procedure; a rod of core glass is placed inside of a tube of cladding glass.

All these processes are based on thermal chemical vapor reactions that form oxides. These oxides are deposited as layers of glass particles called soot, which is deposited on the outer rotating rod or inside glass tube to produce the preforms. Starting materials are solutions of O_2 mixed with $SiCl_4$, $GeCl_4$, 4 $POCl_3$, or gaseous BCl_3. These starting liquids are evaporated within an oxygen stream at a high temperature to produce silicon dioxide (SiO_2) and other oxides. These oxides are known as dopants. Chemical reactions proceed as follows:

$$SiCl_4 + O_2 \rightarrow SiO_2 + 2\ Cl_2$$

$$GeCl_4 + O_2 \rightarrow GeO_2 + 2\ Cl_2$$

$$4\ POCl_3 + 3\ O_2 \rightarrow 2\ P_2O_5 + 6\ Cl_2$$

$$4\ BCl_3 + 3\ O_2 \rightarrow 2\ B_2O_3 + 6\ Cl_2$$

Silicon dioxide (SiO_2), or pure silica, is usually obtained in the form of small (submicron) particles called soot. This soot is deposited on the target rod or tube.

The deposition of the silica soot, layer upon layer, forms a homogeneous transparent material. The manufacturer can control the exact amount of dopant added to each layer, thus controlling the refractive index profile. For example, germanium dioxide (GeO_2) and phosphorus pentoxide (P_2O_5) increase the refractive index of glass, while boron oxide (B_2O_3) decreases it. Changing composition of the mixture during the process influences the refractive index profile of the preform. To change the value of a cladding's refractive index, some dopants are used. For example, fluorine (F) is used to decrease the cladding's refractive index in a depressed cladding material.

The vapor process produces extremely pure material whose characteristics are under the absolute control of the manufacturer. The preforms prepared in the vapor deposition processes are explained in the following sections.

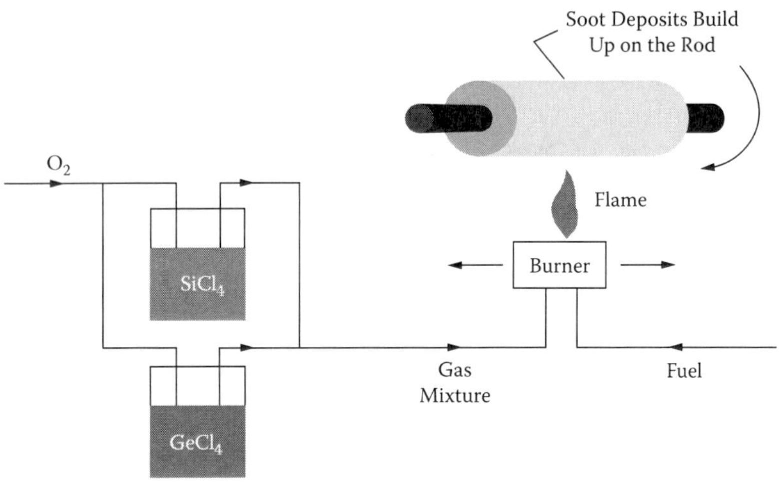

FIGURE 5.8 Outside vapor deposition.

5.12 OUTSIDE VAPOR DEPOSITION (OVD)

This was the first successful mass fabrication process producing preforms used by the fiber drawing method. The outside vapor deposition (OVD) process, also called the soot process, was developed by Corning Company in 1972. This process consists of four phases: lay-down, consolidation, drawing, and measurement. During the lay-down phase, the materials that make up the core and cladding are vapor deposited around the rotating target rod. The result of this process is a soot preform. The refractive index profile and fiber geometry are formed during this phase, as shown in Figure 5.8.

In the consolidation phase, the target rod is removed and the soot preform is placed inside a consolidation furnace. Here the soot preform is consolidated into a solid, clear glass preform and the center hole is closed. During consolidation, a drying gas flows through the preform to remove residual moisture.

In the drawing method, the preform is attached to a precision feed mechanism that feeds it into a furnace at a controlled speed. The drawing method produces a fiber at the required diameter. Later, a color jacket layer will be applied. Diameter measurement and control equipment are also constantly checking the core, cladding, and jacket for diameter sizes and quality.

Finally, in the measurement phase, each fiber reel is tested for compliance with the fiber characteristics given in the data sheets. Specifically, the fiber is tested for strength, attenuation, and dimensional characteristics. Fibers are also tested for bandwidth, numerical aperture (NA), dispersion, and cutoff wavelength.

5.13 VAPOR AXIAL DEPOSITION (VAD)

The vapor axial deposition (VAD) is another form of outside deposition. This method was developed in 1977 by Japanese scientists. Figure 5.9 illustrates the VAD process. Depositing the silica particles, which were obtained from a reaction among gases in a heated zone, occurs at the bottom end of a target, or seed rod, that rotates and

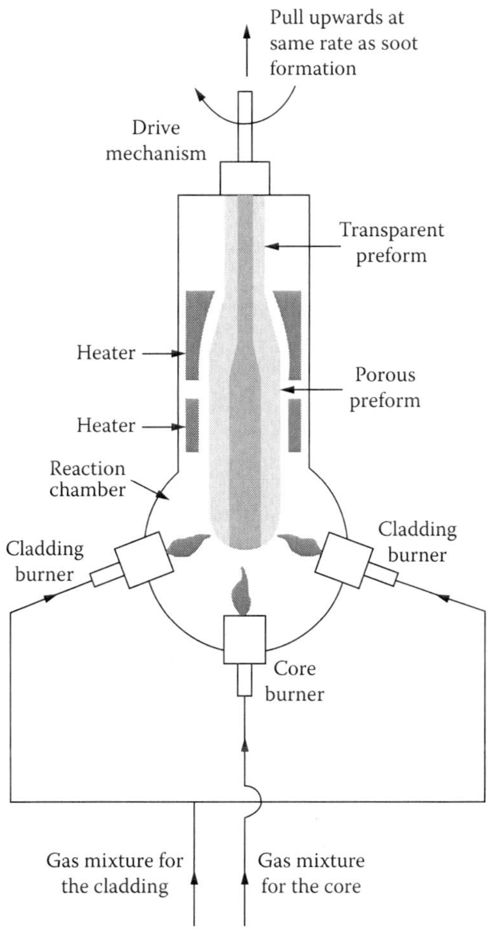

FIGURE 5.9 Vapor axial deposition.

moves upward. This deposition forms a porous preform, the upper end of which is heated in a ring furnace to produce a silica preform. The drawing and measurement steps are similar to those of the other deposition processes. The VAD process does not involve a central hole. The profile of the refractive index is formed by using many burners, a technique that allows the manufacturer to change the direction of the flow of a specific gas mixture. The VAD process produces both step-index and graded-index fiber profiles, resulting when the deposited particle density varies due to temperature gradients produced in the plane perpendicular to the core axis.

5.14 MODIFIED CHEMICAL VAPOR DEPOSITION (MCVD)

This process was developed by Bell Laboratories in 1974 and has been widely accepted for the production of graded-index fiber. In the beginning, there was a process called inside vapor deposition (IVD). This process was significantly improved

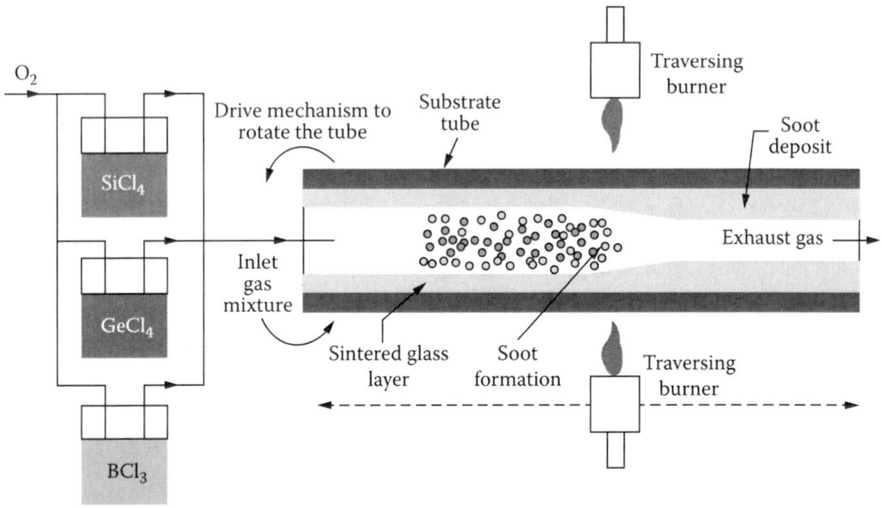

FIGURE 5.10 Modified chemical vapor deposition.

and renamed modified chemical vapor deposition (MCVD). This is the major process used in fiber production in the world. It is illustrated in Figure 5.10.

A mixture of $SiCl_4$, $GeCl_4$, $POCl_3$, O_2, and H_2 gases flows through the inside of the tube, while a heat source surrounds the outside of the tube. The heat source converts the gases into snowlike, high surface area silica soot inside the tube, as shown in Figure 5.10. The soot deposits on the tube downstream of the flame. The burner traverses along the outside of the tube, creating the fine soot particles and sintering the soot into a thin layer of doped glass on the inside of the quartz tube. After the layer is deposited, the mixture of reactive gases is changed and the burner is brought back to the starting position. The above step is repeated and a subsequent layer is deposited. This process is continued, layer by layer, to construct the complex core structure in the optical fiber. Varying the concentration of dopants from layer to layer changes the refractive index, creating a graded-index profile. Once the glass is deposited, the tube is collapsed into a solid rod called a preform. The preform manufactured on the MCVD lathe is heated and drawn down to the standard diameter of 125 microns. Each preform generates many kilometers of fiber. This operation is performed on a draw tower. The tower has a furnace at the top to melt the silica preform. Gauges are used to measure and control the diameter of the glass fiber to within submicrons as the fiber is pulled from the preform. Very fine control of the profile can be obtained by this technique. These MCVD layers are designed to be much thinner than the wavelength of light traveling down the fiber. An acrylate coating is applied during the draw process, which protects the pristine silica fiber from the environment.

After the fiber is manufactured, each spool of fiber is tested to validate the product to meet strict industry and internal specifications. These tests include the measurement of mechanical strength, geometric properties, and optical properties. One key test for laser-optimized multimode fibers is high-resolution differential mode delay (HRDMD).

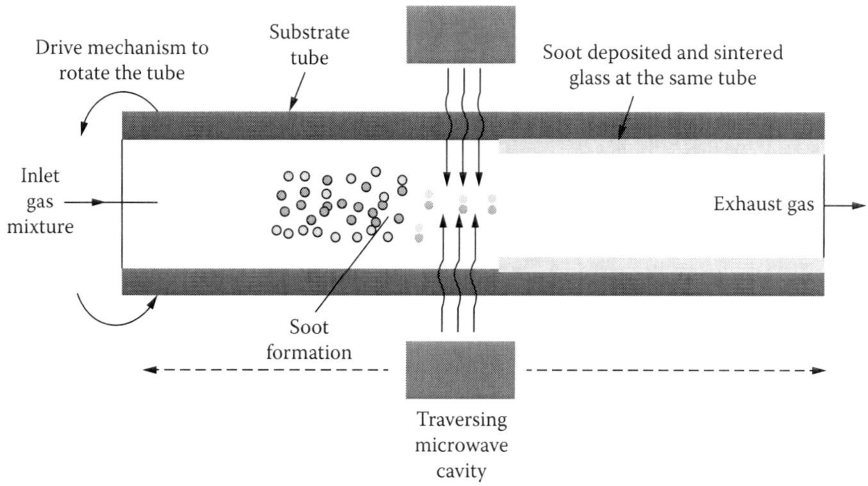

FIGURE 5.11 Plasma-activated chemical vapor deposition (PCVD).

5.15 PLASMA CHEMICAL VAPOR DEPOSITION (PCVD)

The plasma-activated chemical vapor deposition (PCVD) process was developed in 1975 by Phillips Research Laboratories, a Dutch consumer electronics and telecommunications company. The PCVD process is very similar to the MCVD process. Instead of heating the outside of the silica tube, the energy source is provided by a high-power microwave to form ionized gas-plasma inside the silica tube. Figure 5.11 illustrates the PCVD process. Nonisothermal plasma in the microwave frequency range is used instead of a torch or flame. The plasma makes the reaction proceed at about 1,000 to 1,200°C. This results in very thin layers deposited inside the tube. Although this method allows growing layers at relatively low temperatures, the deposition rate is rather slow in comparison to other methods. This process can produce large preforms capable of producing a few hundred kilometers of fiber.

5.16 FIBER DRAWING

Optical fibers are obtained by drawing from the preform at high temperature. The drawing process must be integrated with the jacket coating process to avoid contamination of the fiber surface. These two processes are shown in Figure 5.12. The preform is heated in a furnace to a molten state. The preform is attached to a precision feed mechanism that pushes the preform at the proper speed into the furnace at a high temperature. The drawing process is very precise and controlled continuously to check the diameter of the fiber. Diameter drift cannot exceed 0.1%. The filament fiber passes through a series of coating and jacket applicators and curing processes, depending on the customer's requirements. The outside diameter of the jacket is measured and monitored for defects. The fiber, in its final shape, is pulled down and wound on a winding drum or spool.

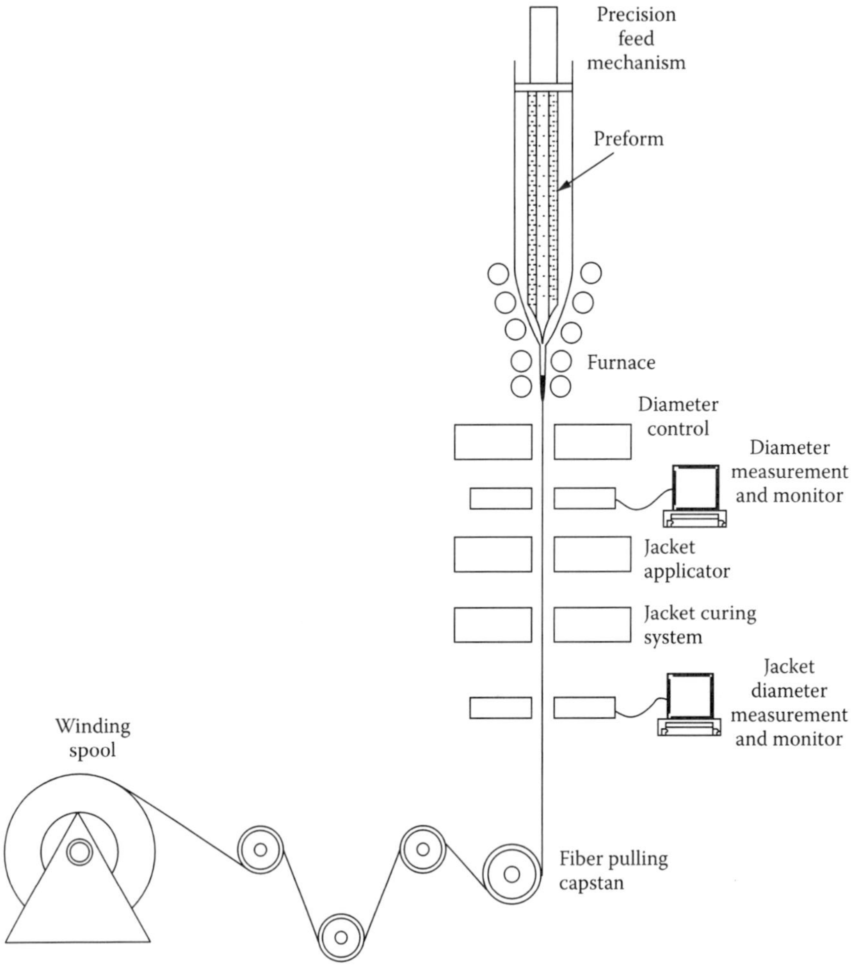

FIGURE 5.12 Fiber drawing and jacket coating.

5.17 PHOTONIC CRYSTAL FIBER CABLES

Recently, there has been a dramatic increase of interest in photonic crystal fiber (PCF), also called microstructure fiber or holey fiber, which features an array of airholes running along the fiber length, as shown in Figure 5.13. PCFs use a microstructured cladding region with airholes to guide light in a pure silica core, giving rise to novel functionalities. The PCFs are divided into two classes: photonic bandgap fibers associated with the photonic bandgap structures and index-guiding holey fibers based on the modified total internal reflection. The latter form their cores by filling one or several airholes in the array with fiber material. This fiber exhibits many unique properties of light guidance, such as single-mode operation over a wide range of wavelengths, highly adjustable effective mode area and nonlinearity, anomalous dispersion at visible and near-infrared wavelengths, and high birefringence. It

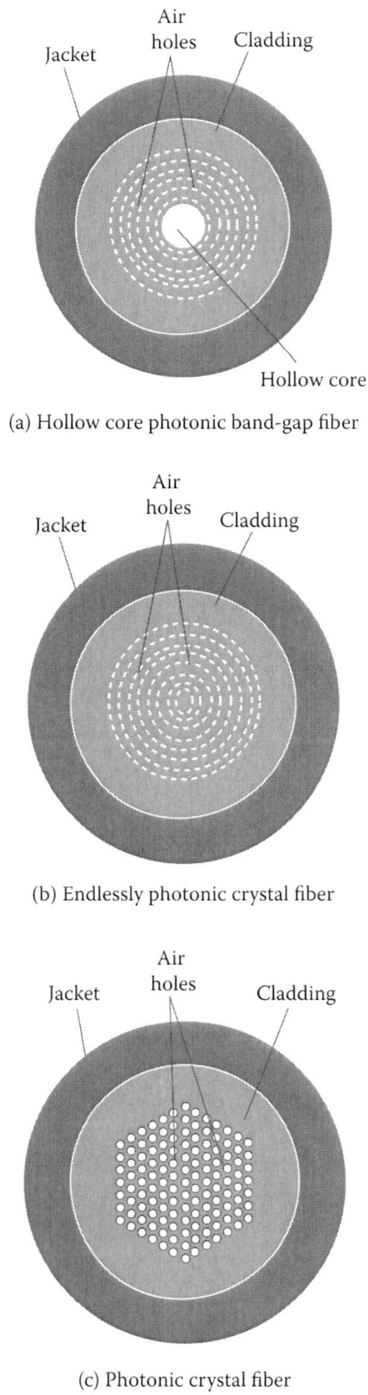

(a) Hollow core photonic band-gap fiber

(b) Endlessly photonic crystal fiber

(c) Photonic crystal fiber

FIGURE 5.13 Cross sections of the photonic crystal fiber cables.

is capable of high-capacity transmission over transoceanic distances, and extremely high-efficiency transmission over a wide range of wavelengths.

Photonic crystal fiber keeps light confined to a hollow core that is about five times thinner than a human hair. The fiber cross section is an air core of about 15 microns in diameter, surrounded by a web-like structure of glass and air. Photonic crystals work on the principle of refraction, or the bending of light as it passes from one material to another. Light travels through the core of the material just as it travels through the middle of hollow fiber optic lines. But photonic crystal is better than a reflective coating at keeping light from scattering or being absorbed, which keeps signals stronger over longer distances. PCFs are widely used in telecommunications, to filter light signals, represent data in quantum communications, implement fiber-based polarization splitters, and transmit exotic wavelengths in the infrared (IR) or ultraviolet (UV) range, for applications in medical equipment and sensors, diode-pumped Nd^{3+} laser sources, and measurements in scientific experiments.

Figure 5.14 illustrates the stack-and-draw process, the most widely used method for fabricating photonic crystal fiber. Silica capillary tubes are stacked by hand to make a preform. The preform has a diameter of 20 to 40 mm. The core is embedded by omitting several capillaries from the center of the stack. Typically, several hundred capillaries are stacked in a close-packed array and inserted into a jacketing tube to create a fiber preform. This preform is drawn into fiber in two stages, adding a second jacketing tube before the final draw. The extra jacket enables the creation of fibers with a standard outer diameter, while providing independent control over the photonic bandgap pitch, and hence the operating wavelength. The preform is drawn into fiber, as shown in Figure 23.8. The preform is attached to a precision feed that moves it into the furnace at a proper speed to soften the silica to a diameter of 2 to 4 mm. The drawing fiber passes through high-precision diameter control and is monitored by imaging or x-ray to detect any nonhomogeneity or bubbles in the drawing fiber. It is then inserted into a silica sleeve tube and drawn down again to a fiber of typically 125 μm in diameter. After this stage, the diameter is continuously monitored by an accurate measurement and monitoring device. The fiber is then covered by a colored jacket layer. Again, the jacket layer undergoes diameter control and monitoring. The end of the fiber cable is attached to a rotating spool, which turns steadily. The fiber cable is then tested for attenuation in dB per kilometer, for dispersion, and any other requirements specified by the customer. Drawn photonic crystal fiber lengths of a few kilometers are typical.

Photonic crystal fibers are available in different types, such as highly nonlinear, double cladding high numerical aperture, large mode field area, and polarization maintaining. These types have different mode field areas, number of holes, and working wavelengths. Some types of photonic crystal fibers have inner and outer cladding layers that are used in highly polarization-maintaining cables.

5.18 MICROSTRUCTURE FIBER CABLES

Microstructure fibers (MOFs) can be used to create a supercontinuum (SC). This is a broadening of the narrow laser spectrum into a broad continuous spectrum with many of the properties of coherent light. SCs have applications in frequency

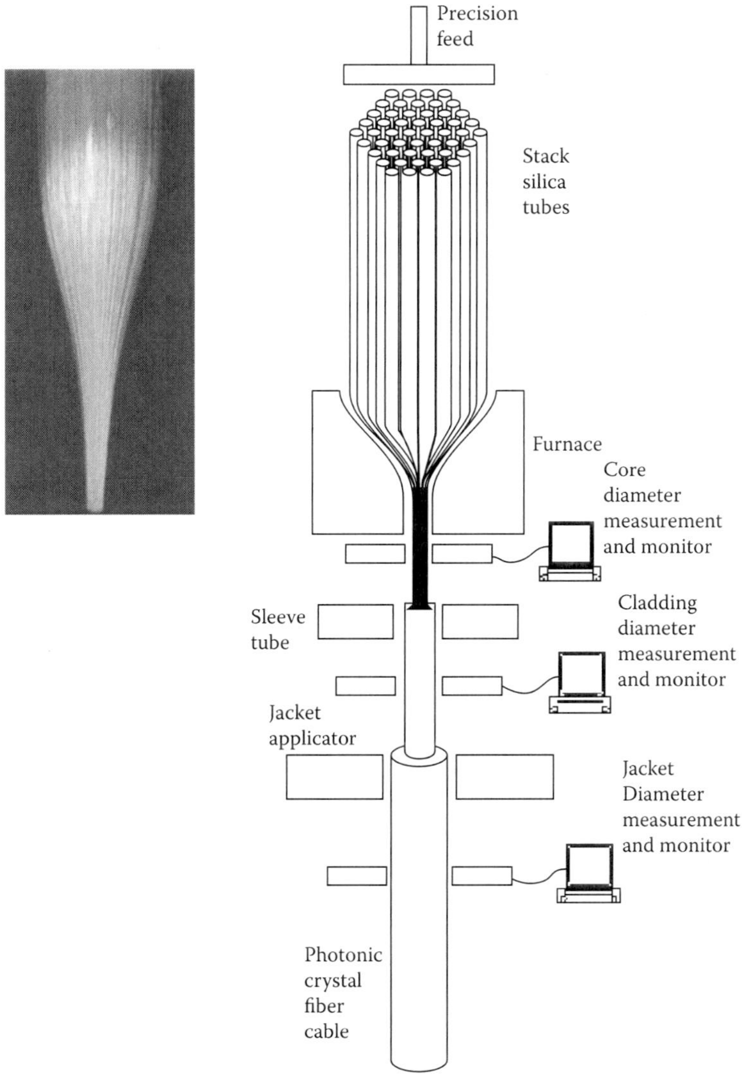

FIGURE 5.14 In the stack-and-draw process for fabricating PCFs.

metrology and optical coherence tomography. The addition of Bragg gratings to the SC light enhances the supercontinuum spectrum, and alters the dispersion profile, thus making gratings an effective tool in tailoring supercontinuum sources.

A grapefruit fiber is one type of microstructure fiber. In the grapefruit microstructure fiber, the lower-order modes are confined by the airholes and are sensitive to surrounding media. Figure 5.15 illustrates a cross section of the grapefruit microstructure fiber. After the addition of long period gratings (LPGs) fiber, the higher-order modes interact with the holes and with the material they may be infused with. The manufacturing process of this fiber is similar to the manufacturing process of

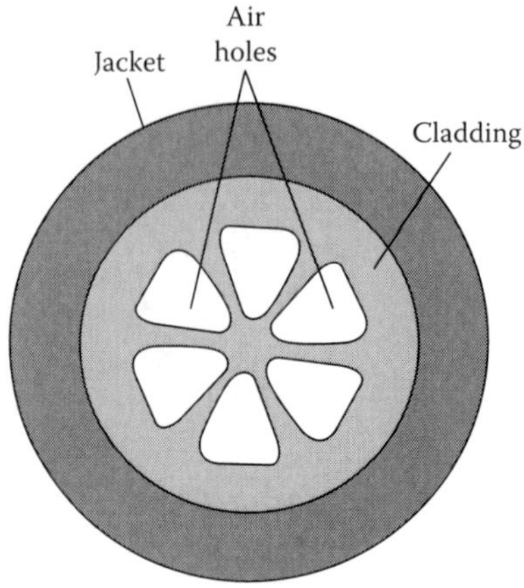

FIGURE 5.15 A cross section of the grapefruit microstructure fiber.

the crystal fibers. This fiber has the same manufacturing process as that of holey fibers. These processes are explained in the above sections.

5.19 POLYMER HOLEY-FIBER CABLES

The polymer holey fiber has the same cross section as the photonics crystal fibers. A polymer fiber is fabricated by first drilling holes into a polymethylmethacrylate (PMMA) preform in which the holes drill down to an intermediate preform. The holes of this intermediate preform are filled with a solution of the dye rhodamine 6G, which permeates the PMMA. The intermediate preform is heated to drive out the solvent molecules and to lock the dye molecules into the structure. Finally, the preform is drawn into a fiber with a 600 μm outer diameter and an 18 μm core diameter. Polymer fibers can be used in replacing the silica fibers in limited applications.

5.20 FIBER CABLE INSPECTION AND HANDLING

This section presents the knowledge to the employees to inspect, clean, and handle fiber optic cables. Figure 5.16 shows the fiber optic cable end preparation and cleaning kit. All safety procedures and regulations regarding the use of fiber optic cables must be followed. The following cases will be performed:

1. Fiber optic cable, 250 μm outside diameter
2. Safety goggles
3. Clean room wipers
4. Finger cots
5. Rubber gloves
6. Cleaver
7. Fiber optic strippers
8. Scissors
9. Cleaner (denatured ethanol)

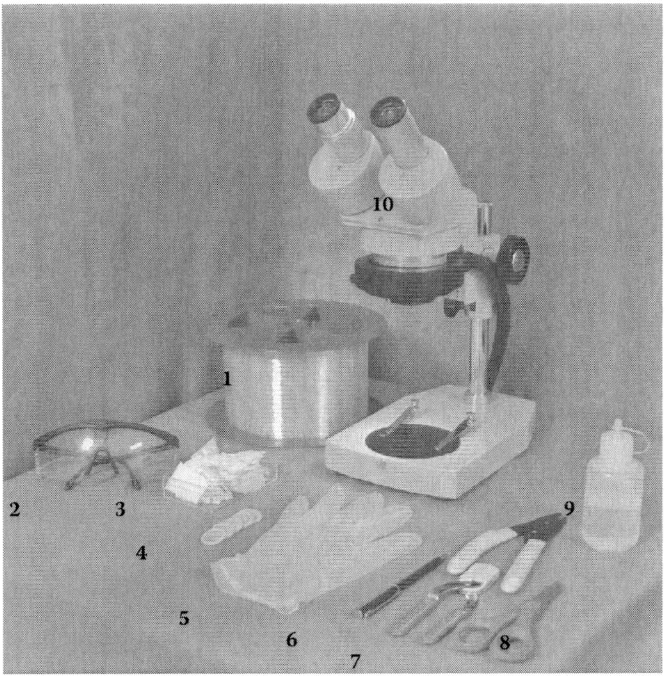

FIGURE 5.16 Fiber optic cable inspection, handling, and end preparation kit.

10. Microscope with a minimum of 10× magnification
11. Thermal cleaver
12. Mechanical cleaver

5.20.1 Case A: Fiber Cable Inspection and Handling

The employees will be able to inspect, clean, and handle fiber optic cables. A full, detailed procedure is presented in the apparatus setup section.

5.20.1.1 Fiber Optic Cable Defect Types

Types of fiber optic cable defects are:

1. Missing jacket: The fiber optic cable jacket is removed, damaged, cracked, or has signs of abrasion.
2. Pinch: The fiber optic cable has an indentation, but there are no breaks in the jacket.
3. Delamination: There is a separation between the fiber optic cable jacket and cladding or a change in the cladding color caused by heat or chemicals, but there are no breaks in the jacket.
4. Contamination: There is a contaminant or any foreign material present on the outer surface of the fiber optic cable jacket.

5.20.1.2 Fiber Optic Cable Inspection

The most common fiber optic cable inspection procedures to find fiber optic cable defects are:

1. Visual inspection: The jacket of the fiber optic cable should be inspected using an illuminated magnifying aid.
2. Touch inspection: The jacket of the fiber optic cable should be inspected using your fingers. To perform touch inspection of the cable jacket, start from one end of the fiber optic cable, gently squeeze the fiber optic cable between the tips of the thumb and index finger, and then slide the fiber optic cable, pulling gently by hand in one direction, to the other end.
3. Combined inspection: Fiber optic cable should be inspected using an illuminated magnifying aid and your fingers to perform the combined (visual and touch) inspection. The visual inspection procedure can be performed under a magnifying aid.
4. Inspection under a microscope: Fiber optic cables should be inspected under a powerful microscope (20×), as shown in Figure 5.17, to find any small crack or damage on the fiber optic cable jacket that cannot be diagnosed by the above inspection procedures.
5. Test inspection: This procedure is an expensive and precise inspection. The fiber optic cable should be tested by sending a signal from a light source and measuring the signal output power. This procedure determines the loss and location of any light leakage from the fiber optic cable jacket.

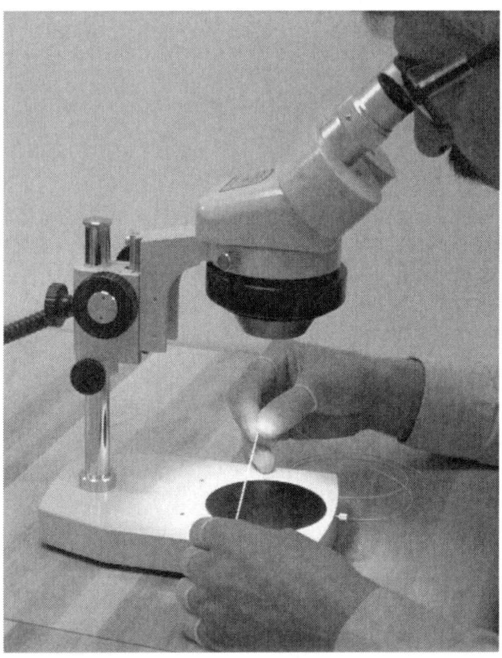

FIGURE 5.17 Fiber optic cable inspection.

5.20.1.3 Fiber Optic Cable Cleaning

The following steps are necessary in the fiber optic cable cleaning process:

1. The fiber optic cable should be inspected and kept clean during the manufacturing process.
2. If any type of contamination is found on the fiber optic cable, follow the recommended cleaning procedure. Each type of fiber optic cable has a cleaning procedure using a specific cleaning solvent. Check the fiber optic cable cleaning process before applying any cleaning solvent.
3. Use a swab or fiber optic cable cleaning pad dampened with solvent to remove the foreign material. Finger cots should be used when handling solvents. Gently rub the swab/pad along the fiber optic cable, in one direction only, to remove the contamination, as shown in Figure 5.18.
4. When cleaning do not use aggressive rubbing to remove the contamination. Do not use fingernails or any other hard surface objects to scrape off the contamination.
5. If unable to remove the foreign material using the above methods, reject the fiber optic cable.

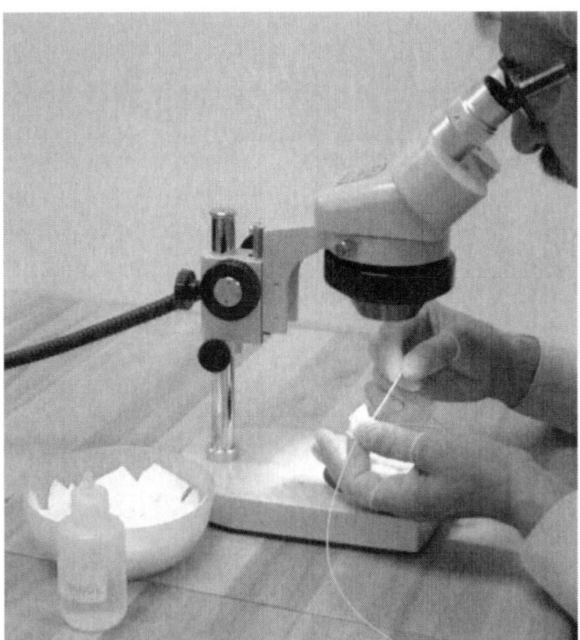

FIGURE 5.18 Fiber optic cable cleaning.

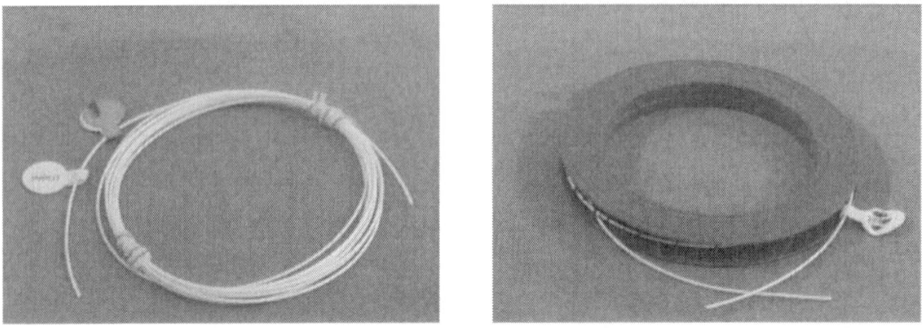

(a) Coiled fiber (b) Coiled fiber on a rubber ring

FIGURE 5.19 Fiber optic cable handling.

5.20.1.4 Fiber Optic Cable Handling

When handling fiber optic cable, observe the following rules:

1. Do not allow the fiber optic cable to rest on the floor.
2. Do not allow the fiber optic cable to get crushed or pinched.
3. Do not allow the fiber optic cable to drop directly over sharp edges, such as a table edge, tools, or handling trays.
4. Do not allow the fiber optic cable to come in contact with hot surfaces, such as a soldering iron, hot air pencil, or hot handling trays from an oven.
5. Do not allow anything to be placed on top of the fiber optic cable during handling.
6. Do not permit any kind of macro-bends to be present in a fiber optic cable. Bends increase attenuation and decrease the tensile strength of the fiber optic cable.
7. The fiber optic cable should be coiled, as shown in Figure 5.19(a). The coil diameter depends on the fiber optic cable diameter and type. The fiber cable can also be coiled on a rubber ring, as shown in Figure 5.19(b). The size of the rubber ring depends on the fiber optic cable diameter and type.
8. Always separate paper from the fiber optic cable. Put paper in a plastic bag when handling the fiber optic cable.

5.20.2 Case B: Fiber Cable Ends Preparation

The employees will also be able to prepare fiber optic cable ends. A full, detailed procedure is presented in the apparatus setup section.

5.20.2.1 Manual Stripping Procedure

1. Hold and open the hand stripper with one hand only.
2. Hold the fiber optic cable very tightly between the thumb and forefinger. Place the stripper on the fiber optic cable, making sure to insert the fiber

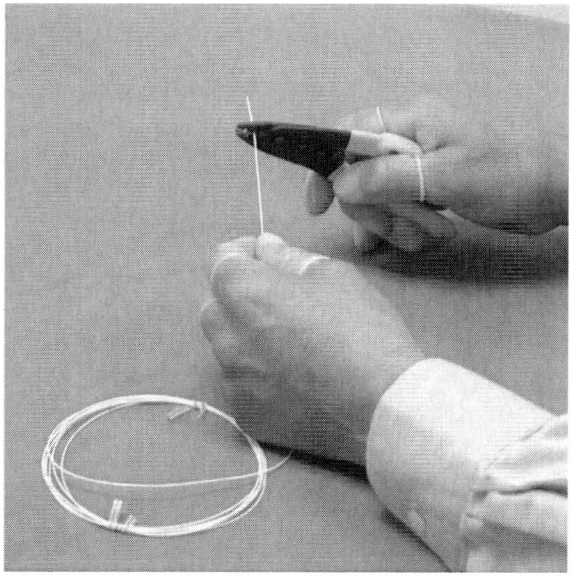

FIGURE 5.20 Manual stripping.

optic cable through the V's in heads, as shown in Figure 5.20. The stripper should be perpendicular to the fiber optic cable. Approximately 1 in. of the fiber optic cable should pass through to the other side of the stripper.

3. Gently squeeze the handles until the stripper closes completely. Keep the handles in this position.
4. While holding the fiber optic cable tightly, pull the stripper along the length of the fiber optic cable, as straight as possible, toward the fiber optic cable end.
5. Always keep the hand stripper clean.

5.20.2.2 Thermal Stripping Procedure

The thermal strippers are easy, quick, and safe tools for the stripping of jackets compared to the mechanical strippers, which create a mechanical stress and scratches on the surface of the cladding. The following steps summarize the procedure:

1. Turn on the thermal stripper.
2. Hold and open the thermal stripper with one hand only.
3. Hold the fiber optic cable very tightly between the thumb and forefinger. Insert the fiber cable into the stripper for the stripping length that is required, as shown in Figure 5.21. The stripper should be lined up with the fiber optic cable. Approximately 1 in. length of the fiber optic cable should pass through into the stripper.
4. Gently squeeze the handles until the stripper closes completely. Keep the handles in this position.

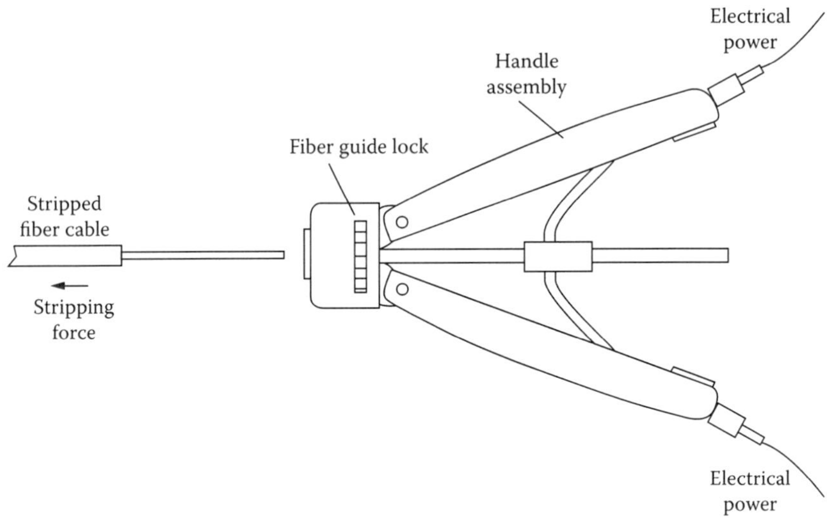

FIGURE 5.21 Thermal stripping.

5. While holding the fiber optic cable tightly, pull the stripper along the length of the fiber optic cable, as straight as possible, away from the fiber.
6. Always keep the thermal stripper clean. Follow the cleaning procedure of the device.

5.20.2.3 Cleaning Procedure after Stripping

1. Take a piece of cotton tissue dampened with denatured ethanol or any cleaning chemical suitable for the fiber.
2. Wrap it around the stripped end of the fiber optic cable.
3. Pull in one direction toward the end.
4. Clean more than once if necessary.

5.20.2.4 Manual Cleaving Procedure

1. Place the stripped length of fiber optic cable firmly on the inside part of your forefinger.
2. Take the cleaver with your other hand, holding it perpendicular to the fiber optic cable. Make a gentle scratch across the cladding surface of the fiber optic cable at a distance of about ½ in. from the end of the fiber optic cable, as shown in Figure 5.22.
3. Use the same tool to break off the fiber optic cable at the scribed mark.

5.20.2.5 Mechanical Cleaving Procedure

A mechanical cleaver is used in most production lines. The mechanical cleaver gives a precise cut at a cleave angle of 90° to the fiber optic cable end. Figure 5.23 shows a mechanical cleaver, which cleaves the fiber optic cable end to a required length; this

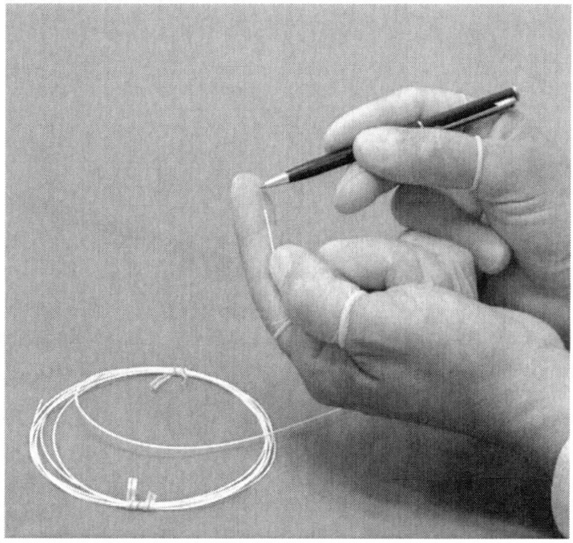

FIGURE 5.22 Manual cleaving.

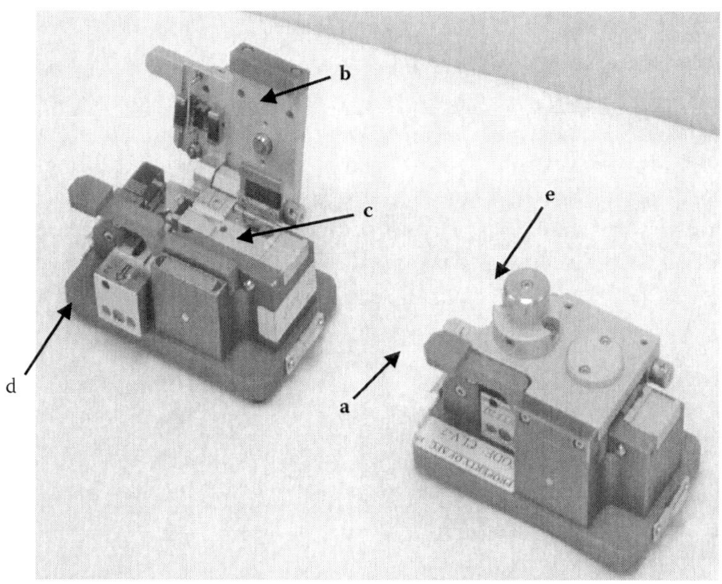

FIGURE 5.23 Mechanical cleaving.

is typical in most manufacturing and test processes. To use the mechanical cleaver perform the following steps:

1. Get a mechanical cleaver ready.
2. Press the red lever (a), as shown in Figure 5.23.
3. Open the top mechanism (b).
4. Open the small cover (c).
5. Place the stripped fiber optic cable at the required cleave length on the disk blade on the block (d). This is where the cleaving will take place.
6. Close the small cover (c) and top mechanism (b).
7. Push block (d) in the direction of the arrow, to cleave the fiber.
8. Press knob (e), to break the fiber.
9. Press the red lever (a), open the top mechanism (b), and open the small cover (c).
10. Remove the fiber optic cable.
11. The fiber optic cable is cleaved automatically to the required cleave length.

5.21 MANUFACTURING FIBER CABLE ASSEMBLY

The fiber tube assembly is typically the first step when manufacturing an optical device. The fiber tube assembly provides a means to handle, position, and glue the fiber optic cable to an optical component, such as a GRIN lens, Y-coupler, or beam-splitter. The fiber tube assembly is manufactured from single or double fiber optic cable and a glass tube, as shown in Figure 5.24. The end of the fiber cable can be prepared as explained above. The fiber optic cable is attached to the tube using two types of epoxy. A hard type of epoxy fills the front part of the tube and a soft type of epoxy fills the other end. Air bubbles in the epoxy or cracks in the tube will introduce insertion loss, especially under temperature variations. Fiber optic cable must also not be cracked or stressed, as losses will increase. The fiber tube assembly end is polished at an angle of 2, 6, 8, or 12° according to the design requirements where the fiber tube assembly will be used. The assembly end must be polished precisely and carefully to maintain a surface free of cracks and scratches. Figure 5.25 shows the fiber tube assemblies ready to be joined to a GRIN lens or an optical device. The fiber tube assembly has a polished end angle matching the polished end angle of the

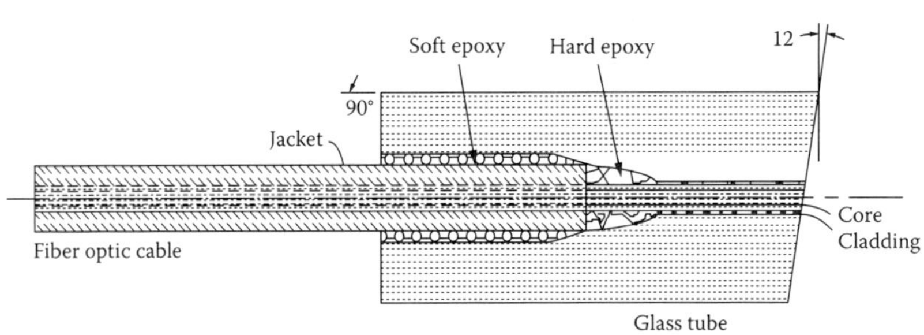

FIGURE 5.24 A fiber tube assembly/single fiber cable.

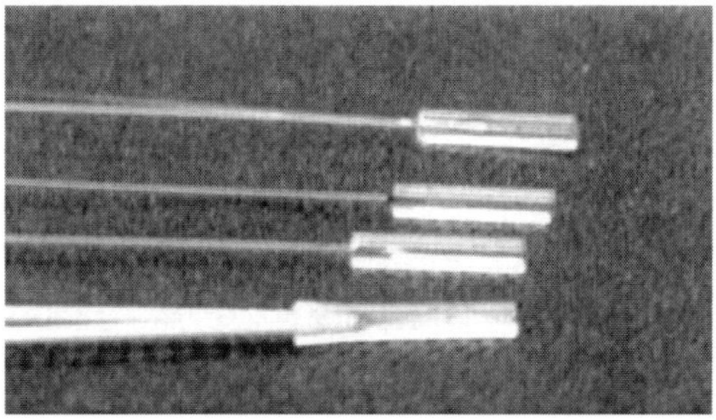

FIGURE 5.25 Fiber tube assemblies.

GRIN lens. An accurate optical alignment is achieved when the fiber tube assembly and the GRIN lens are facing each other in perfect alignment. The ends of the fiber tube assemblies are polished to the required angles using an industrial polishing machine for mass production.

5.22 LIGHT SOURCE COUPLING TO A FIBER CABLE

There are several techniques for coupling a light source to a fiber optic cable. The coupling process can be done during manufacturing of a laser light source, which is connected to a fiber cable at one end and a pigtail connector on the other end, as shown in Figure 5.26. Light-emitting diodes and laser diodes have edge or top surface light emission. Therefore, coupling of a cable to the emitting side of a light source depends on many factors, such as the design and packaging size of the device. There are also similar processes used in coupling a fiber cable to an optical component, such as a prism or a GRIN lens.

For an efficient coupling process, the emitting area of the light source should be equal to or slightly larger than the core of the fiber cable, to ensure that all the light passes through the cable with minimum loss. Figure 5.27(a) shows one of the most

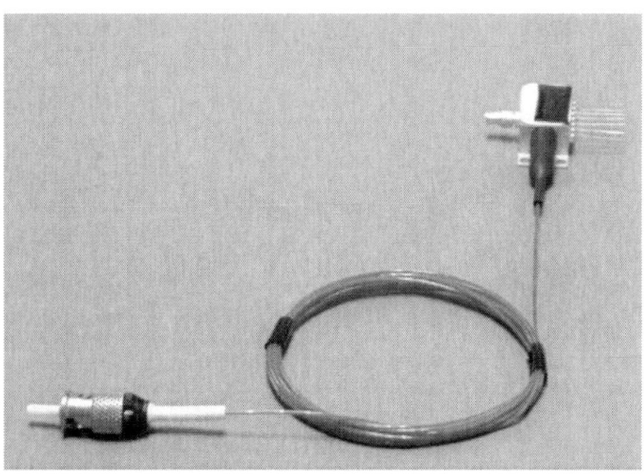

FIGURE 5.26 A light source coupled to a fiber cable.

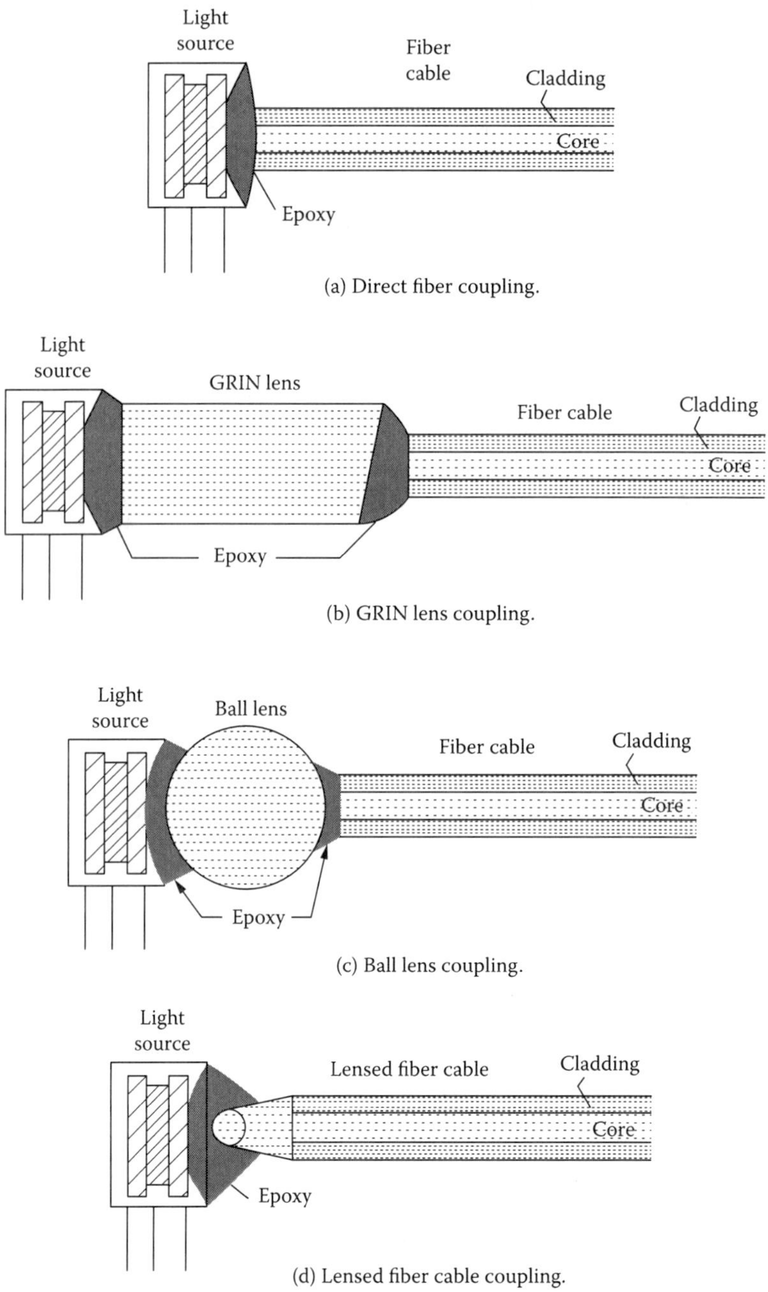

(a) Direct fiber coupling.

(b) GRIN lens coupling.

(c) Ball lens coupling.

(d) Lensed fiber cable coupling.

FIGURE 5.27 Common methods of coupling a light source to a fiber cable.

common coupling processes, called direct coupling. An optical epoxy can be used to bond the fiber cable to the light source. Application of the epoxy and curing process is very important during manufacturing to reduce losses. This method is the cheapest process of coupling, because of the short alignment time and a simple assembly design to enable alignment of the light source to the fiber cable. This method is less efficient than the other coupling methods.

To improve the coupling process, an optical component can be used, such as a lens, GRIN lens, or ball lens, and grating and a prism. A problem arises when the light-emitting area and the fiber cable are not the same size, creating losses. Using a lens that reduces the light beam size in order to match the fiber cable can solve this problem. Figure 5.27(b) illustrates the use of a GRIN lens placed between the light source and the cable. GRIN lenses are commonly used in building many optical fiber devices, such as switches and polarization beamsplitters, because of easy alignment by a fairly skilled employee. A ball lens can also be used in the coupling process, as shown in Figure 5.27(c). Optical devices using ball lenses take more time for alignment to the light source. These devices are more complicated to design and bigger in size than the devices using GRIN lenses. In this case, light coupling suffers high loss because more epoxy is needed to bond the ball lens in position. Figure 5.27(d) illustrates the coupling of a lensed fiber cable to a light source. This arrangement has an easy coupling alignment with low loss and small packaging size.

5.23 LAUNCHING LIGHT CONDITIONS INTO FIBER CABLES

The amount of light carried by each mode in a fiber cable will be determined by light input or light launch conditions. If the angular spread of the rays from the light source is greater than the angular spread that can be accepted by the fiber cable (the NA of light input to the fiber cable is greater than the NA of the fiber), then the radius of the light input is greater than the radius of the fiber cable. This case is called overfilled, as shown in Figure 5.28(a). In this case, a portion of the light source will be launched into the cladding and will be considered light loss. Conversely, when the input light NA is less than the NA of the fiber cable, i.e., the radius of the light input is smaller than the radius of the fiber cable, this is called underfilled, as shown in Figure 5.28(b). These two coupling conditions will yield different attenuations, with the overfilled case having a higher loss than the underfilled case.

There are also other factors that have to be taken into consideration in coupling: condition of the fiber cable end, index of refraction of the epoxy, epoxy curing process, alignment method, type of optical components, and packaging. Similarly, it is necessary to consider some of these factors in any type of connection between two fiber cables. Common factors are the angular alignment, air gap, different numerical apertures or core diameters, and cable end conditions.

5.24 APPLICATIONS OF CONNECTORS AND SPLICES

Connectors and splices make optical and mechanical connections between two fiber cables. It is easy to connect and disconnect a cable with a connector from another

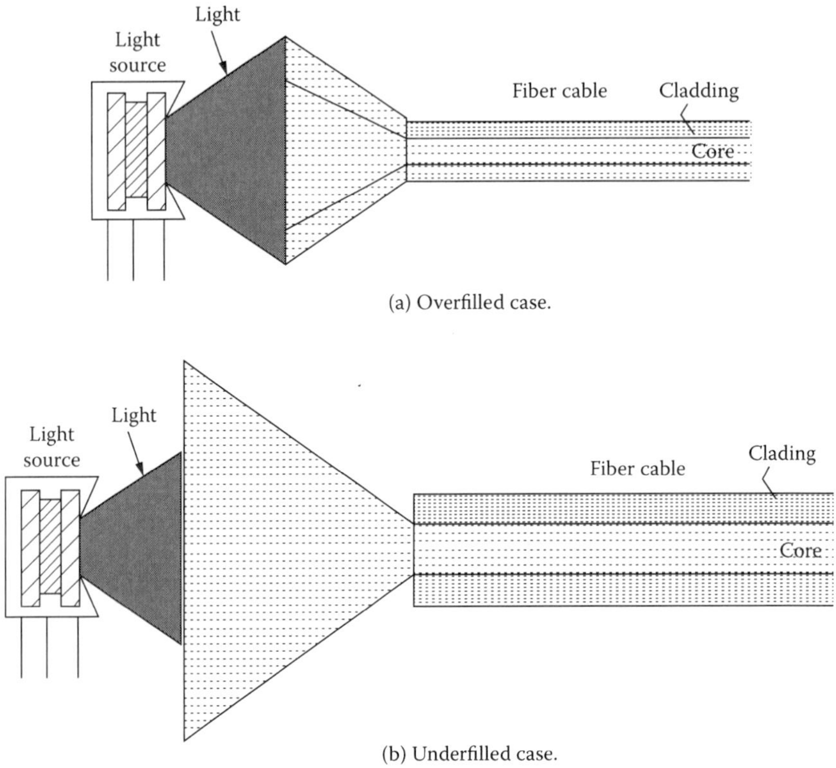

FIGURE 5.28 Launching light into a fiber cable.

cable or a device. There are many applications for fiber connectors and splices in fiber systems, including:

1. Connecting between a pair of fiber cables, using connectors or a splice, is an essential part of any fiber system.
2. Interfacing devices to local area networks.
3. Connecting and disconnecting fiber cables to patch panels where signals can be checked and routed in a fiber system.
4. Connecting and splicing may be required on short fiber cables for wiring, testing devices, connecting instruments and devices, and at other intermediate points between transmitters and receivers.
5. Dividing a fiber system into subsystems, which simplifies the selection, installation, testing, and maintenance of fiber systems.
6. Temporarily connecting remote mobile systems and recording equipment in many fiber systems.

5.25 REQUIREMENTS OF CONNECTORS AND SPLICES

It can be very difficult to design a connector or a splice that meets all the requirements. A low-loss connector may be more expensive than a high-loss connector, or it may require relatively expensive application tooling. The lowest losses are desirable, but the other factors clearly influence the selection of the connector or splice.

The following is a list of the most desirable features for fiber connectors or splices that are required by customers and industry:

1. Low loss (insertion and return): The connector or splice causes low loss of optical power across the junction between a pair of fiber cables.
2. Easy installation and use: The connector or splice should be easily and rapidly installed without the need for special tools or extensive training.
3. Repeatability: There should be no variation in power loss. Loss should be consistent whenever a connector is connected, disconnected, and reconnected again, as many times as required.
4. Economical: The connector, splice, and special application tooling should be inexpensive.
5. Compatibility with the environment: The connector or splice should be waterproof and not affected by temperature variations.
6. Mechanical properties: The connector or splice should have high mechanical strength and durability to withstand the application and tension forces.
7. Long life: The connector or splice should be built with a material that has a long life in various applications.

5.26 FIBER CONNECTORS

Fiber connectors are designed to be easily connected and disconnected, as shown in Figure 5.29. Fiber optic cable can be easily connected to a transmitter, receiver, power meters, or another fiber cable. The key optical parameter for a fiber optic

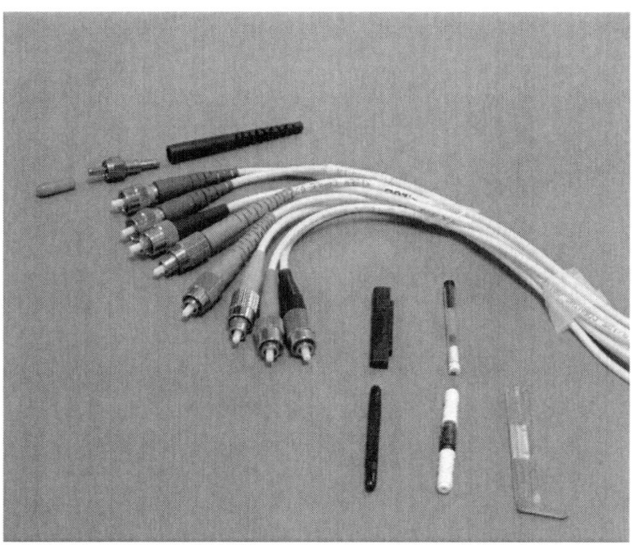

FIGURE 5.29 Fiber optic connectors and splices.

connector is its attenuation. Signal attenuation in connectors is the sum of losses caused by several factors. The major factors are as follows:

1. Overlap of fiber cable cores (also called lateral displacement)
2. Alignment of fiber axes
3. Fiber cable numerical aperture
4. Reflection at the fiber cable junction/interface
5. Connector end polishing
6. Fiber cable spacing
7. Connector end face profiles
8. Insertion loss

5.27 FIBER OPTIC CONNECTOR TYPES

Figure 5.30 shows the most common types of fiber optic connectors. Fiber connectors are unique in that they must make both optical and mechanical connections. Fiber connectors must allow the fiber cables to be precisely aligned to ensure that a connection is robust. Fiber connectors use various methods to achieve solid connections. Some of the types of fiber optic connectors currently in use are listed below:

- Subscriber connectors (SC)
- SC/APC connectors
- FC/PC connectors
- FC/APC connectors
- LC connectors
- MU connectors
- Straight-tip connectors (ST)
- 5685C connectors (duplex SC)

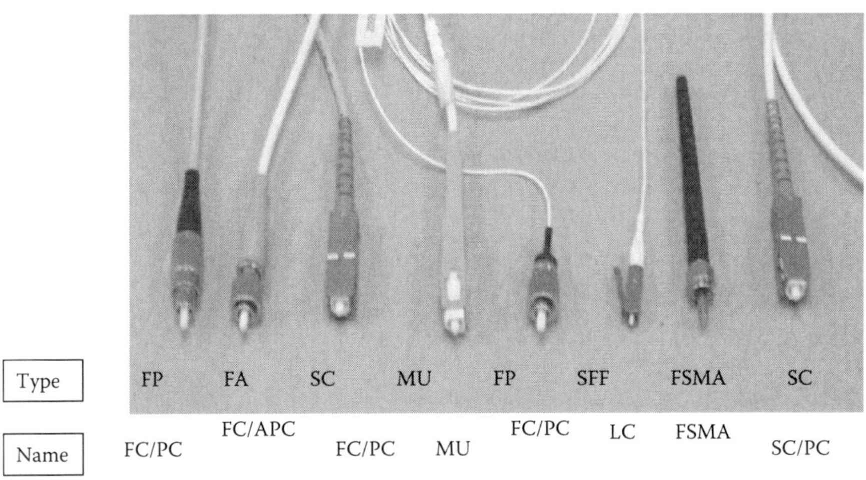

FIGURE 5.30 Fiber optic connector types.

- FDDI connectors (MIC)
- Biconic connectors
- SMA connectors
- Enterprise system connection (ESCON)
- Duplex connectors (ST)
- Polarizing connectors
- MT multifiber connectors
- MT-RJ connectors
- D4 style connectors
- Biconic connectors
- MFS/MPO connectors
- Plastic fiber connectors
- E-2000 diamond
- Fiber optic connectors that self-latch in a push-pull system
- Special connectors

5.28 ADAPTERS FOR DIFFERENT FIBER OPTIC CONNECTOR TYPES

An adapter is a passive device used to join two different types of connectors together. The type of adapter is identified by a nomenclature, such as SC, FC, ST, and 568SC. Hybrid adapters join dissimilar connectors together, such as SC to FA. Figure 5.31 shows examples of some adapters.

5.29 FIBER OPTIC CONNECTOR STRUCTURES

Most fiber optic connectors are built from a ferrule, a connector body, an epoxy material, and a strain relief boot. Most connectors use a ferrule to hold the fiber and provide alignment. The most popular ferrule size is a 2.5 mm diameter, which is standard. Manufacturers offer a few types of ferrules made from different materials, such as ceramic, plastic, and stainless steel.

Connectors may be attached to a device, outlet box, or adapter, by direct connection, by coupling a threaded nut, or by twisting a spring-loaded bayonet socket. The connector body is made from steel, ceramic, or plastic. Epoxy is usually applied to secure the fiber cable end in the connector body. A strain relief boot made from plastic or rubber is used at the junction between the connector body and the fiber cable.

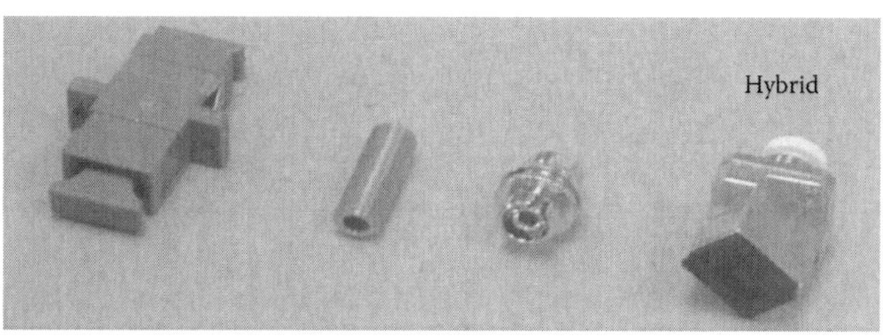

FIGURE 5.31 Fiber optic adapter types.

5.30 FIBER OPTIC CONNECTOR ASSEMBLY TECHNIQUES

The following sections present common assembly techniques that are used in building fiber optic connectors.

5.30.1 Building FSMA Connectors

The most common fiber connector assembly techniques use a fiber cable and a suitable connector. The fiber cable is most often epoxied into the connector. Epoxy provides good tensile strength to the connector to prevent the fiber cable from moving within the connector body, maintaining a good alignment. After the epoxy cures, the ferrule end is polished to a smooth finish by one of the many available procedures. Then the connector undergoes many inspections and test procedures to issue a data sheet for the customer.

The following tools and devices are needed to build FSMA connectors:

1. 2 × 2 ft optical breadboard
2. HeNe laser light source and power supply
3. Laser light sensor
4. Laser light power meter
5. Laser mount assembly
6. 20× microscope objective lens
7. Lens/fiber cable holder/positioner assembly
8. Fiber cable holder/positioner assembly
9. Hardware assembly (clamps, posts, screw kits, screwdriver kits, lens/fiber cable holder/positioner, sundry positioners, brass fiber cable holders, fiber cable holder/positioner, etc.)
10. Fiber cable end preparation procedure and kit, and cleaning kit, as explained in Chapter 6, end preparation section
11. Connector holder/positioner assembly
12. Black/white card and cardholder
13. Water spray bottle
14. 50%/50% mixing epoxy
15. Fiber optic cable, 900 μm diameter, 500 m long
16. Polishing disk
17. FSMA connector, as shown in Figure 5.32

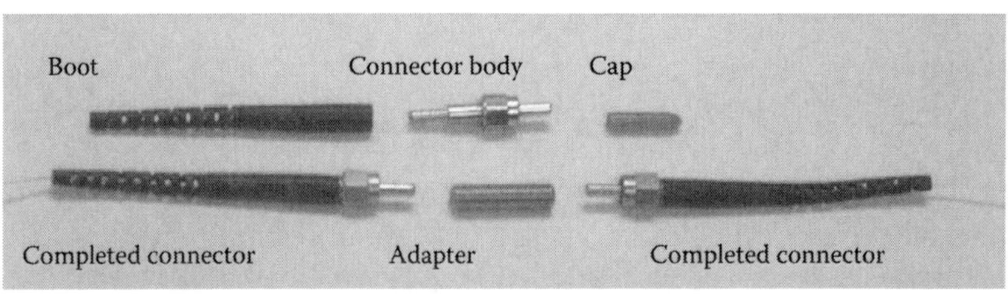

FIGURE 5.32 FSMA connectors.

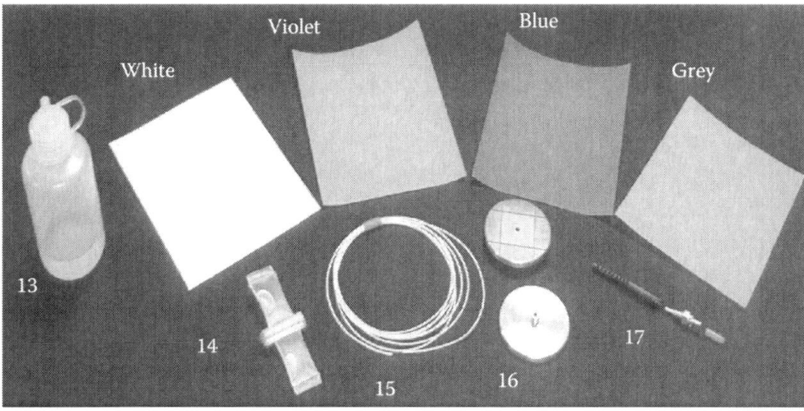

FIGURE 5.33 Connector assembly kit.

18. Polishing pads, as shown in Figure 5.33
    Size 63.0 μm grey color
    Size 9.0 μm blue color
    Size 1.0 μm violet color
    Size 0.3 μm white color
19. Key lock mechanical fiber optic splicing unit, as shown in Figure 5.42
20. Table type mechanical fiber optic splice unit, as shown in Figure 5.43
21. Fusion splicing machine, as shown in Figure 5.46

5.30.1.1 Connector Body

The connector body is a continuation of the ferrule, as shown in Figure 5.32. The connector body accommodates the strain relief boot. The connector is attached to an adapter or a device by a threaded coupling nut.

5.30.1.2 Epoxy and Polish

A fiber cable is often epoxyed to a connector. Epoxy is generally considered to be an undesirable but necessary process in fiber optic termination, because of its curing time. After the epoxy cures, the ferrule end is polished to a smooth finish.

5.30.1.3 Strain Relief Boot

A black polymer strain relief boot shields the junction of the connector body and the fiber cable.

Figure 5.33 shows the connector assembly kit. Building and testing a connector involves the following steps:

1. Cut two 4 ft lengths of fiber cables from the spool.
2. Strip, cleave, and clean the fiber cable ends so that there is about 1 to 1¼ in. of bare fiber extending beyond the jackets. Follow the fiber cable end preparation procedure as explained in Chapter 6.

3. Prepare a connector for the assembly process (the connector consists of two parts: ferrule connector body and strain relief boot).
4. Insert the prepared end of the fiber cable into the connector, until the fiber cable stops and the bare fiber emerges from the ferrule.
5. Prepare the epoxy according to the epoxy manufacturer's preparation procedure.
6. Pull the fiber cable ½ in. backward, and then, with a toothpick or small probe, apply the epoxy on the fiber cable. Push and pull the cable (back and forth), until the epoxy adequately fills the gap between the fiber cable and the ferrule. The quantity of epoxy must be sufficient to support the fiber cable inside the connector body.
7. Insert the fiber cable into the ferrule completely.
8. Using a small probe, place a very small drop of epoxy onto the ferrule face where the fiber cable exits. This will seal off the space between the ferrule face hole and the fiber cable. The size of this drop is important; too large an epoxy drop will extend the polishing time.
9. Set the connector aside for the specified time, to allow the epoxy to fully cure. Curing time depends on the type of epoxy.
10. Insert the fiber cable through the strain relief boot, positioning the boot on the connector body.
11. Cleave the protruding fiber flush with the ferrule face. Remove the protruding piece of fiber and dispose of it in the designated container.
12. Place the size 63.0 μm gray color polishing pad on a flat and clean surface, such as a piece of glass sheet.
13. Screw the connector gently onto the polishing disk until finger-tight. Do not overtighten.
14. Place several small drops of water on the polishing pad. Begin the polishing process by moving the polishing disk with your fingers in a figure 8 motion, as shown in Figure 5.34.

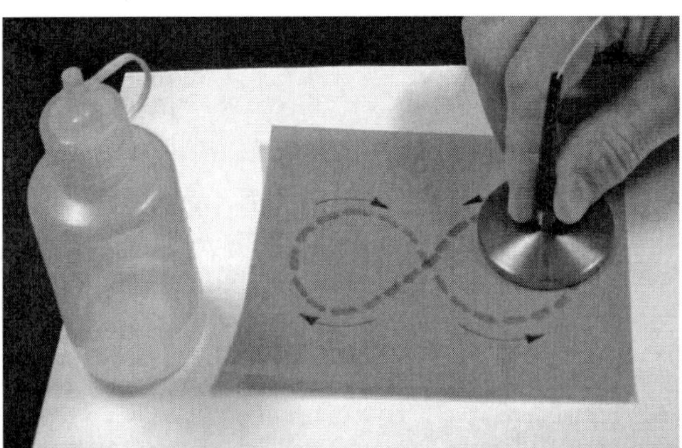

FIGURE 5.34 Polishing connector by hand.

15. Polish until the epoxy bead and excess fiber cladding are removed, and the fiber cable is flush with the ferrule face surface.
16. Apply only light pressure on the polishing disk, using enough water to keep the polishing pad and disk clean; apply constant motion.
17. About 20 to 30 motions should be sufficient to complete the rough polishing.
18. Rinse the polishing disk with water.
19. Inspect the ferrule face using an inspection microscope. Look at the quality of the polished ferrule face.
20. Repeat steps 14 to 19 using polishing pads of sizes 9.0 μm (blue color), 1.0 μm (violet color), and 0.3 μm (white color).
21. About 20 to 30 motions should be enough to achieve a good quality finish on the ferrule face when using the white pad. **Caution:** Do not overpolish.
22. Inspect the finished connector using the inspection microscope.
23. Once the two connectors are completed, test the connectors to measure the connection loss, which is calculated using Equation 5.1:

$$\text{Loss}(\text{dB}) = -10\ \log_{10}\left(\frac{P_{\text{out}}}{P_{\text{in}}}\right) \tag{5.1}$$

5.30.1.3.1 Testing Connection Loss in Two Connectors

1. Figure 5.35 shows the apparatus setup for testing connection loss in two connectors.
2. Follow the instruction of the fiber optic cable loss measurements and calculations explained in Chapter 6.

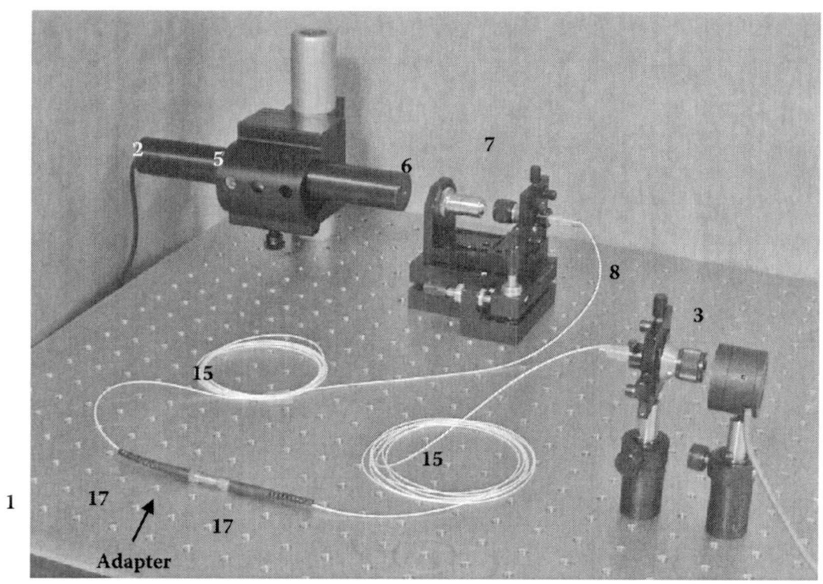

FIGURE 5.35 Mated connectors setup test.

3. Measure the laser input power into the fiber cable.
4. Couple the laser beam output to the fiber cable input.
5. Carefully align the laser with the lens/fiber cable holder/positioner, so that the maximum amount of the laser beam is entering the core of the fiber cable.
6. Check to ensure that you have a circular output from the first optic connector. Direct the output toward the center of the laser sensor.
7. Measure the laser output power from the first optic connector.
8. Screw the two optic connectors into the in-line adapter, with one connector on each side, as shown in Figure 5.35.
9. Measure the laser output power from the second fiber cable end, as shown in Figure 5.35.
10. Determine the connection assembly loss from the data. Use a working formula for fiber optic connector loss.
11. Mate and unmate the connectors three times to check the repeatability of the connection loss figures.

5.30.2 Hot-Melt Connectors

Hot-melt connectors are widely used in North American telecommunication systems. The hot-melt connectors use preloaded epoxy so that external mixing is not required. The prepared end of the fiber optic cable is inserted into the connector ferrule, as shown in Figure 5.36. The cable (with the connector inserted) is loaded onto the connector holder and placed in an oven for a few minutes. The oven will soften the epoxy around the fiber cable and cure the epoxy at the same time. The curing time is dependent on the type of epoxy. The end of the connector is then polished to a smooth finish. The polishing can be achieved by hand or by an industrial

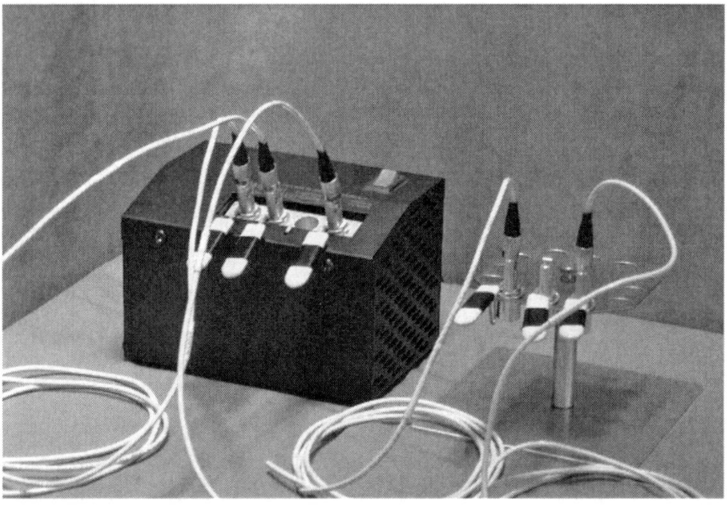

FIGURE 5.36 Hot-melt connectors assembly.

polishing machine. When such connectors are assembled in the field, a portable hand polisher is used.

5.30.3 Epoxyless Connectors

Epoxyless connectors, called crimp connectors, have been widely used for quick cable connections in telecommunication systems. When the connector is crimped, an insert compresses around the fiber cable. A front clamp on the bare fiber cable and a rear compression clamp add a higher clamping force on the fiber cable buffer coating to provide the necessary tensile strength. Special gripping tools are used in the assembly of the epoxyless connectors. The end of the fiber connector is polished to a smooth finish using a portable hand polisher before the connector is assembled in the field. The main advantage of an epoxyless connector is the speed of assembly. Some customers will tolerate a slightly higher loss to achieve a fast, easy termination. The epoxyless approach is a technology that is not limited to one connector type.

5.30.4 Automated Polishing

All the fiber optic connector-polishing machines are designed for accuracy, easy setup, and production efficiency. The polishing pressure, speed, and duration can be adjusted to meet exact requirements. These machines precisely polish the ends of fiber optic connectors in a repeatable and reliable manner. Polishing machines are available for dry or wet polishing process.

5.30.5 Fluid Jet Polishing

Fluid jet polishing (FJP) is another technique for shaping and polishing small surface areas of complicated optics made of brittle materials. This technique uses a fluid jet system to guide premixed slurry, at low pressures, onto the optical surface being machined. The surface is altered by the erosive effect of the abrasive particles in the stream.

5.30.6 Fiber Optic Connector Cleaning

Contamination of connector ends can occur from something as simple as dust particles or fingerprints that can reduce light propagation through the fiber cable. This will degrade device performance, causing data error and loss. To avoid this, it is common practice to clean fiber connectors prior to assembly and testing.

There are three major components of the fiber optic connector system that users must consider when cleaning, mating, and testing fiber optic connectors: the adapter split sleeve, the outer diameter of the ferrule, and the tip of the ferrule. There are many techniques for cleaning connectors, either wet or dry, by hand with recommended cleaning chemicals, or with automated machines. Follow the cleaning procedure for each fiber connector type. Do not use the same procedure for other types of connectors. Cleaning standards for fiber optic connectors promise savings in time and cost.

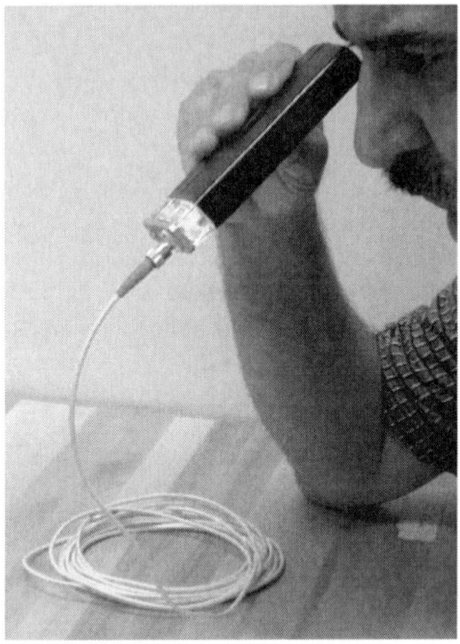

FIGURE 5.37 Connector end inspection with a view scope.

5.30.7 Connector Testing

There are many testing instruments available for testing connectors. Testing instruments range from a simple view scope to a sophisticated system. The condition of the end of the ferrule after the polishing process is usually inspected using simple instruments, as shown in Figure 5.37. This procedure is adequate for inspecting a connector built and polished in the field.

Handheld devices can measure the losses, optical powers, light sources, etc. The basic test measures the attenuation of the fiber cable with connectors, by comparing the power through the fiber cable to that of a known reference fiber cable. The power through the fiber cable under test is measured in absolute units. The power through the reference fiber cable is also measured. Figure 5.38 illustrates a connector test setup.

Using sophisticated systems for testing connectors saves time and cost in industrial production. These systems are very accurate. Each connector type refers to a standard test. The preferred test methods are compliant with the commercial building telecommunications cabling standard, TIA/EIA-568-B.1. The ANSI/TIA/EIA standards group developed intermateability standards for several connector types to ensure compatibility among manufacturers.

5.31 FIBER SPLICING

The splicing process joins fiber optic cable ends permanently. In general, a splice has a lower loss than a connector. Splices are typically used to join lengths of cable for outside applications. Splices may be incorporated into lengths of fiber optic cable

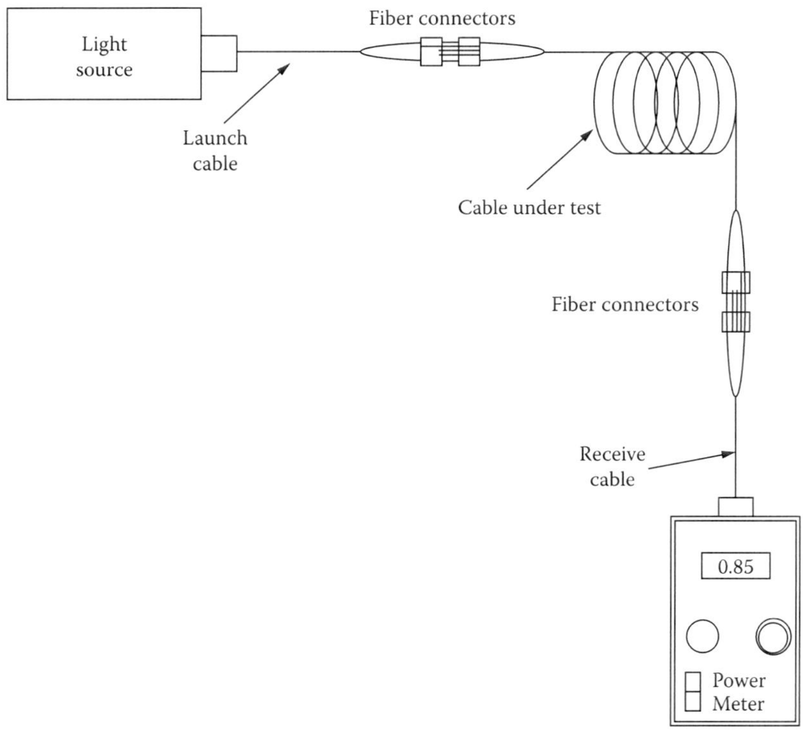

FIGURE 5.38 Connector testing setup.

or housed in indoor/outdoor splice boxes, whereas connectors are typically found in patch panels or attached to equipment at fiber cable interfaces. The sources of loss, described in Section 5.26, are also applicable to splices. There are two types of splices: mechanical and fusion.

5.31.1 Mechanical Splicing

Mechanical splices join two fiber cable ends together both optically and mechanically, by clamping them within a common structure. In general, mechanical splicing requires less expensive equipment. However, higher consumable costs are experienced. Figure 5.39 shows the most common types of fiber optic mechanical splices. A few important types of mechanical splices are listed below:

- Table type splices
- Key lock splices
- Fiber lock splices
- Twist lock splices
- Fastomeric splices
- Capillary splices
- Rotary or polished ferrule splices
- V-groove splices
- Elastomeric splices
- Finger splices
- Inner lock splices

Many other types are available.

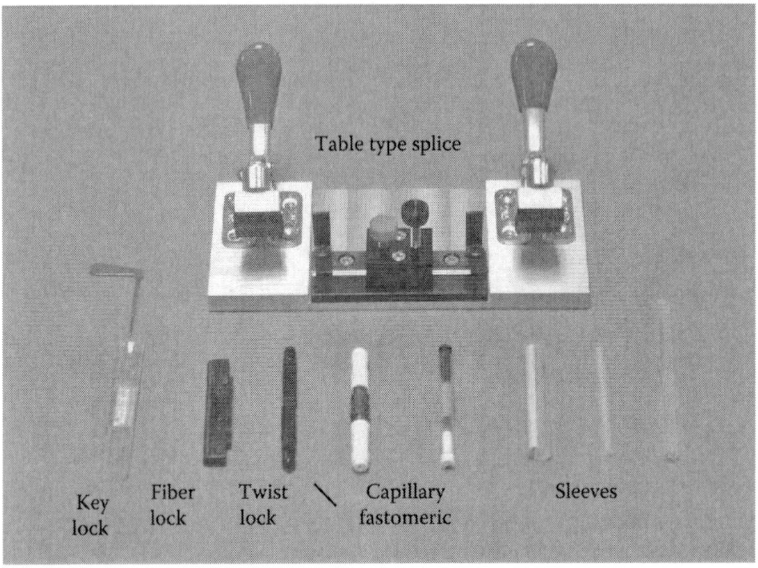

FIGURE 5.39 Some of the fiber optic splice types.

Some mechanical and fusion splices are used with one type of splice closure. Figure 5.40 shows small and large splice closures. Splice closures are standard pieces of hardware in the telecommunication industry for protecting fiber optic cable splices. Splices are protected mechanically and environmentally within the sealed closure. Splice closures are waterproof. Water is kept out by using nonflowing gel under permanent compression. They are suitable for indoor, outdoor, and underground cable system installation. There are small and large closures available for different applications.

Splice trays are designed to hold fusion and mechanical splices, as shown in Figure 5.41. They are available in different sizes. Fusion and mechanical splices are held in a specially designed splice organizer and splice holder, respectively. They are not sealed off environmentally. These trays are installed in a wall-mounted fiber splice cabinet in a communication system.

The following sections explain temporary mechanical splices that are used in different cases of the testing.

5.31.2 Key Lock Mechanical Fiber Optic Splices

Figure 5.42 shows the key lock mechanical fiber optic splice, which is commonly used to quickly mate and unmate fiber optic cables. It is made from a U-shaped metal part, which is covered by a transparent plastic body with two holes on each end. The prepared ends of the fiber cables are made longer than half of the length of the metal part. The fiber cable is inserted in the center hole. When the key is inserted in the second hole toward the edge of the splice and turned 90°, the metal part opens and then one fiber cable end can be easily inserted. This operation can be repeated on

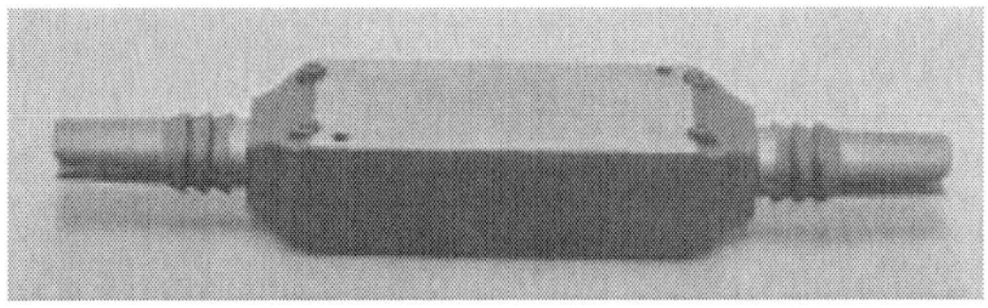

(a) A small closure

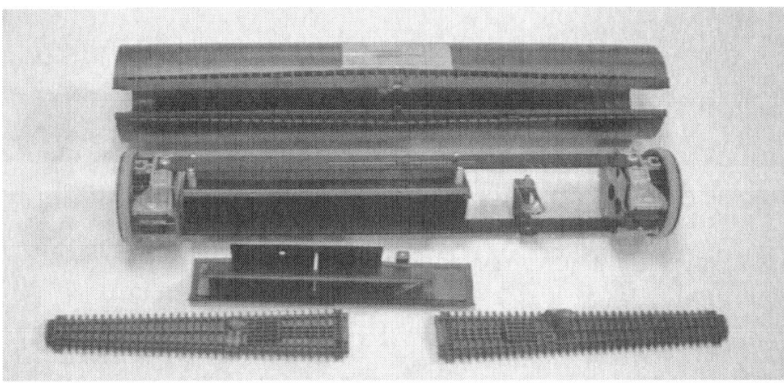

(b) A large closure

FIGURE 5.40 Splice closures.

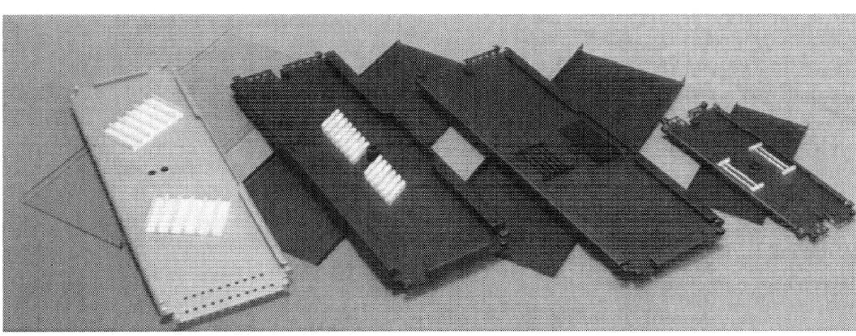

FIGURE 5.41 Splice trays.

the other side to insert the second fiber cable. This type of splice provides a quick and easy way of joining two fiber cables with low signal loss. It may be used to temporarily or permanently connect fiber cables, wavelength division multiplexing components, and other fiber optic elements.

5.31.3 Table Type Mechanical Fiber Optic Splices

Figure 5.43 shows a custom-made mechanical splice, which is used for quick mating and unmating of connections. This splice works like any other mechanical splice. The fiber cable ends are prepared and inserted into the midpoint of the block

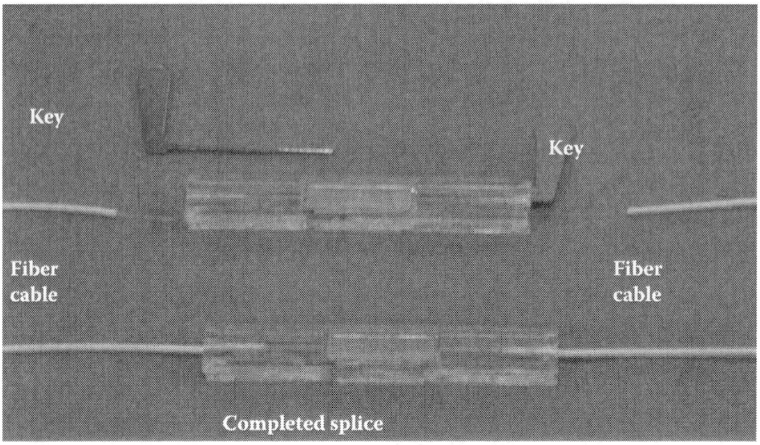

FIGURE 5.42 Key lock mechanical fiber optic splice.

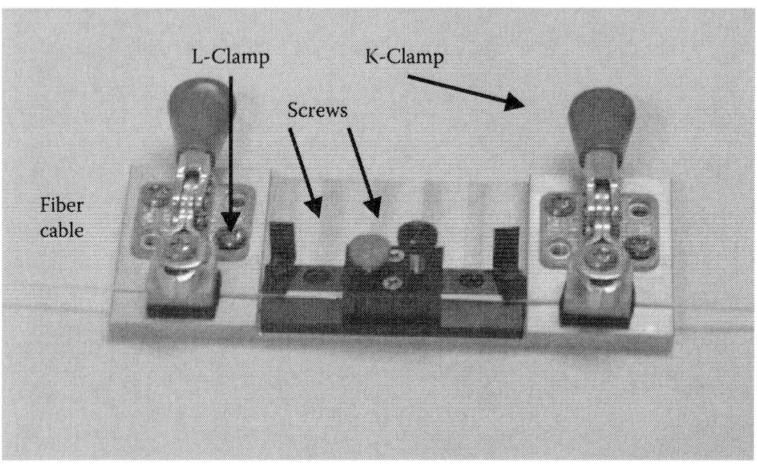

FIGURE 5.43 Table type mechanical fiber optic splice.

assembly. Screws are tightened to align the fiber cables on both sides. L-clamps and K-clamps are placed in position to secure the fiber cables on both sides. Most fiber optic companies use this kind of mechanical splice for quickly mating and unmating during manufacturing and testing processes. The splice loss associated with these instruments is acceptable by industry standards.

5.31.4 Fusion Splices

Fusion splicing is performed by placing the tips of two fiber cables together and heating them by a fast electrical fusion process, so that they melt into one piece. Fusion splices automatically align the two fiber cable ends and apply a spark across the tips to fuse them. Fusion splicers also include instrumentation to test the splice quality

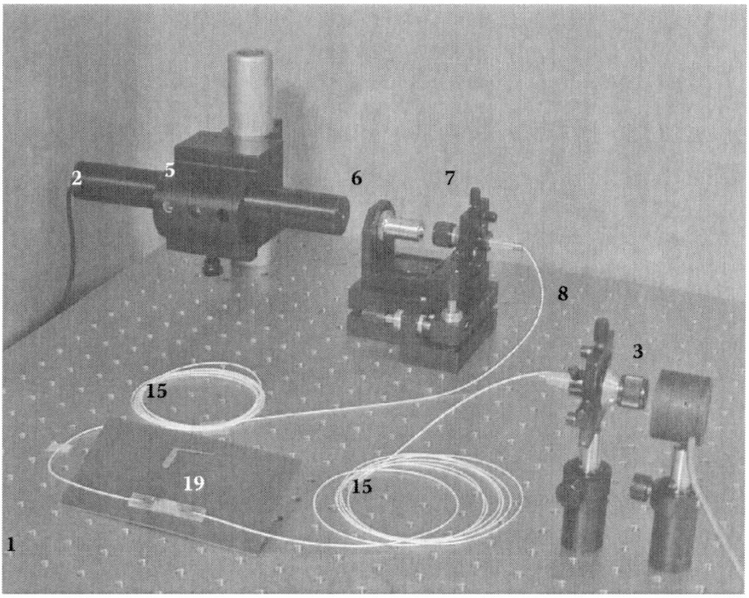

FIGURE 5.44 Key lock mechanical splicing setup test.

and display optical parameters pertaining to the join. A fusion splice is shown in Figure 5.46. When the fusion splice is completed, a cylindrical fusion protector is placed over the splice location. Fiber fusion protectors are made from metal or polymer, and they are applied to ensure mechanical strength and environmental protection. Some types of fusion splice protectors (sleeves), as shown in Figure 5.39, are designed for use in place of the heat shrink method for fast, easy, and reliable permanent installations. Fusion splices provide lower loss than mechanical splices.

5.31.5 Splice Testing

Attenuation can be measured as the splice is being performed. Many fusion splicers come with instrumentation that measures the loss as the fiber cables are being aligned, and test the loss when the splice is completed.

5.31.5.1 Testing Connection Loss Using a Key Lock Mechanical Splice

1. Figure 5.44 shows the apparatus setup for testing connection loss in a mechanical splice. Prepare a key lock mechanical splice for the fiber cable connection process.
2. Repeat steps 2 to 7 from case b.
3. Insert the key fully into the edge hole and turn it 90° to the open position.
4. Insert the prepared first fiber cable output end into the center hole on the side of the key lock mechanical splice. The fiber cable end should be at the midpoint of the metal part of the mechanical splice. Then turn the key to the closed position to secure the fiber cable in the mechanical splice. Remove the key.

5. Repeat step 4 and gently insert the second fiber cable end into the other side of the mechanical splice. The two fiber cables should meet approximately at the center of the metal part. Make sure that the fiber cable ends are as close (face-to-face) as possible without an air gap between them. If the mechanical splice shines during the test, the light is escaping because the two ends are not close enough.
6. Measure the laser output power from the second fiber cable, as shown in Figure 5.44.
7. Determine the key lock mechanical splicing loss from the collected data.
8. Repeat the mechanical splicing process three times to check the repeatability of the connection loss figures.

5.31.5.2 Testing Connection Loss in a Table Type Mechanical Splice

1. Figure 5.45 shows the apparatus setup for testing connection loss in a table type mechanical splice.
2. Prepare the table type mechanical splice by cleaning and oiling it to be ready for the connection process.
3. Repeat the fiber cable ends as explained above.
4. Insert the first prepared fiber cable output end into the table type mechanical splice to the midpoint of the black block assembly on one side.
5. Tighten the screw on the side where the first fiber cable was inserted, to secure the first fiber cable in the splice.
6. Rotate the L-clamp and clamp down the K-clamp into position, to secure the first fiber cable.

FIGURE 5.45 Table type mechanical splicing setup test.

7. Repeat steps 4 to 6 to insert the second fiber cable into the other side of the splicer.
8. Measure the laser power at the second fiber cable output.
9. Determine the connection loss from the data.
10. Repeat the mechanical splicing process three times to check the repeatability of the connection loss figures.

5.31.5.2.1 Testing Connection Loss in a Fusion Splice

1. Figure 5.46 shows the apparatus setup for testing connection loss in a fusion splicer machine.
2. Prepare a fusion splicer machine for the fiber cables connection process.
3. Repeat the steps in the mechanical splice for the fiber cables connection process.
4. Turn on the fusion splicer.
5. Insert one prepared end of each fiber cable into the bonder.
6. Make sure that the fiber cable ends are lined up straight, to ensure proper bonding.
7. Look through the window of the fusion splicer to inspect the positioning of the two fiber cables; ensure that they are lined up and flush. Perform a fine alignment for the two fiber cable ends.
8. Push the fusion key to start the fusion process.
9. The fusion splicer connects the two fiber cables permanently. When the splicing is completed, remove the fiber cable.
10. Add a fiber optic fusion protector (sleeve) to the fusion location on the fiber cable.
11. The fusion splicer has a display panel and can display the splice test results and other optical parameters, as shown in Figure 5.46.

FIGURE 5.46 Fusion splice setup test.

12. Set up the fusion splice test, as shown in Figure 5.46.
13. Measure the laser power at the second fiber cable output.
14. Determine the fusion splice connection loss from the data.
15. Find the connection loss from the data that is provided by the fusion splicer machine.

6 Fiber Optic Cable Types and Installations

6.1 INTRODUCTION

Fiber optic cables are significantly different from copper cables. Fiber optic cables transmit data through very small core diameters over long distances at the speed of light. Fiber optic cables come in a wide variety of configurations. Important considerations in any cable installation and operation are the bending radius, tensile strength, ruggedness, durability, flexibility, environmental conditions such as temperature extremes, and even appearance. Due to fiber optic cables' light weight and extreme flexibility, they are more easily installed than their copper counterparts. They are easy to handle and can be pulled through conduit and piping systems over long distances using various installation techniques. The minimum bend radius and maximum tensile load allowed on a fiber optic cable are critical during and after installation. A tensile load causes attenuation and may ultimately crack the fiber optic core. The tensile load allowed during installation is higher than the permissible loads during operation. The minimum bend radius allowed during installation is larger than the bend radius allowed during operation.

This chapter presents the types of fiber optic cables and their methods of installation. Testing fiber cables after installation is an important aspect for any communication system. Some hardware used in telecommunication systems is presented in detail. This chapter also introduces the student to fiber cable installation and the testing of the fiber cable after installation is complete.

6.2 FIBER OPTIC CABLE TYPES AND APPLICATIONS

There are many fiber optic cable types and designs depending on the application. Fiber cables can be classified into the following categories: fiber optic cables for indoor applications, outdoor applications, and indoor/outdoor applications. The following sections explain each category in detail.

6.2.1 Indoor Fiber Optic Cable Types and Applications

Indoor cables (inside cable plant) are generally installed and operated in a controlled, stable environment. The cables must perform with minimal loss. Other factors that generate losses, such as environmental and mechanical stresses during and after installation, can cause failure. Outdoor cables (outside cable plant) have more factors that affect their performance. Indoor fiber cables are divided into the following types:

1. Simplex cables contain a single fiber cable, as shown in Figure 6.1. They are used for connections within equipment. They have a thicker outer jacket, which makes the cable easier to handle and also adds mechanical protection.

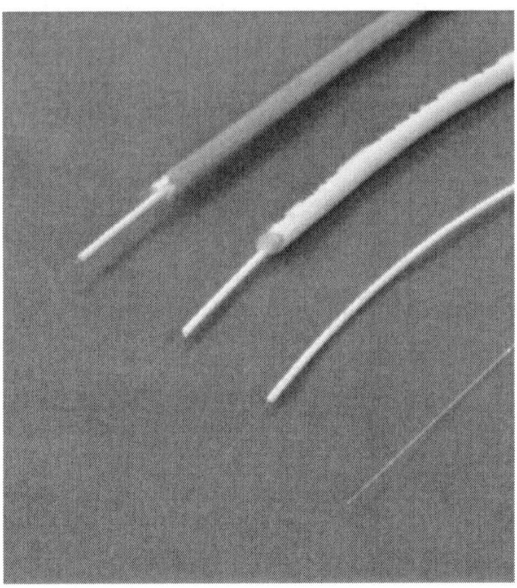

FIGURE 6.1 Simplex fiber cables.

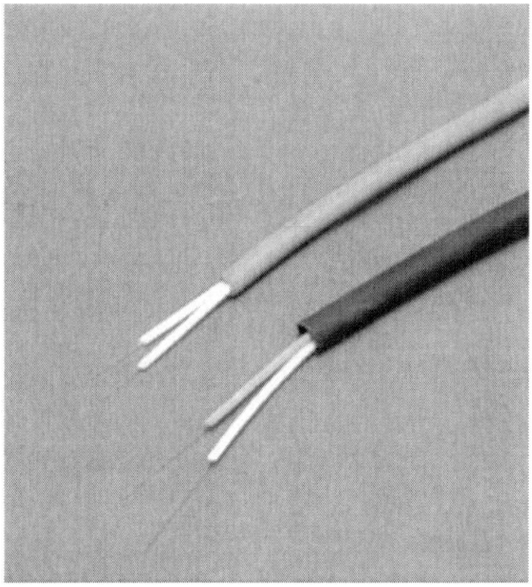

FIGURE 6.2 A duplex fiber cable.

2. Duplex cables contain two tight-buffered fiber cables inside a single jacket, as shown in Figure 6.2. They are used in equipment interconnections within workstations, test equipment, hubs and routers, etc.
3. Multifiber cables contain more than two fiber cables in one jacket, as shown in Figure 6.3. They have anywhere from three to several hundred fiber cables, which are all color coded. They are used in vertical and horizontal fiber cable distributions between floors and in a telecommunication room.
4. Light cables, heavy cables, and plenum cables are application dependent.
5. Breakout cables have several individual simplex cables.

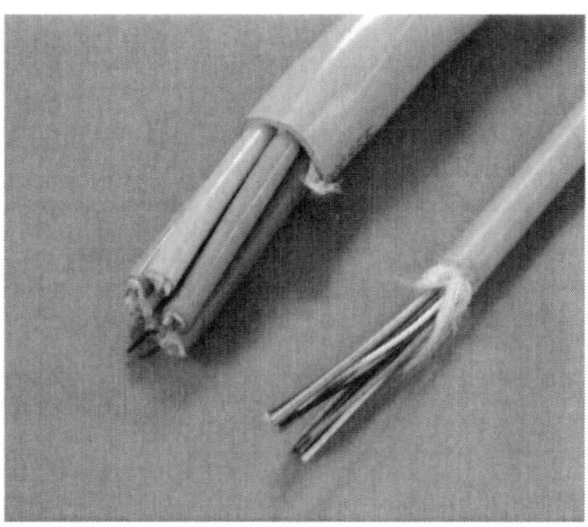

FIGURE 6.3 Multifiber cables.

6.2.2 Outdoor Fiber Optic Cable Types and Applications

Outdoor cables (outside cable plant) must withstand a variety of environmental and mechanical stresses during and after installation. The cables must perform with minimal losses over a wide range of temperature and humidity changes. The cable must also have waterproof capabilities, have strength to endure the difficult installation conditions, provide protection against ultraviolet radiation, and provide mechanical protection. There are many types and designs depending on the manufacturer and application. Outdoor fiber cables are divided into the following types:

1. Overhead cables may be strung from telephone poles or along power lines.
2. Direct burial cables are placed directly in a trench dug in the ground and then covered by soil.
3. Indirect burial cables are placed inside a duct or conduit system.
4. Submarine cables are laid underwater.

6.2.3 Indoor/Outdoor Fiber Optic Cable Types and Applications

There are many types of fiber cables that are used for indoor and outdoor applications. These cables use materials that enable them to pass the flame-retardant requirements of the indoor applications and provide reliable waterproof performance for outdoor applications. They are able to withstand difficult installation conditions and temperature variations. They are widely used within the building and between buildings in campus applications.

6.2.4 Other Fiber Optic Cable Types and Applications

One of the commonly used fiber cables is the ribbon cable. Ribbon cables are made of many fibers, which are embedded in a plastic or PVC material in parallel, forming a flat, ribbon-like structure, as shown in Figure 6.4. The ribbon cable carries up to 12 fiber cables in a single ribbon. They are ideal for multifiber connector applications from equipment to a patch panel or as a patch cord.

Some ribbon cables consist of up to 216 fiber cables, as shown in Figure 6.5. The cable consists of a single buffer tube that contains a stack of up to eighteen 12-fiber ribbons wrapped within a water-swellable foam tape and surrounded by a jacket. Dielectric or steel strength members located under the cable jacket provide tensile and mechanical strength. Some ribbon cables provide up to 864 fiber cables in a rugged, compact design to maximize the use of critical duct space. They are the ideal choice to maximize usage of duct or pipe space and get the service running faster than other cable types. They are ideal for outdoor applications and upgraded communication systems.

6.3 FIBER OPTIC CABLE INSTALLATION METHODS

This section describes some of the most common fiber optic cable installation methods for inside and outside plants in local area networks, metropolitan area networks, and wide area networks.

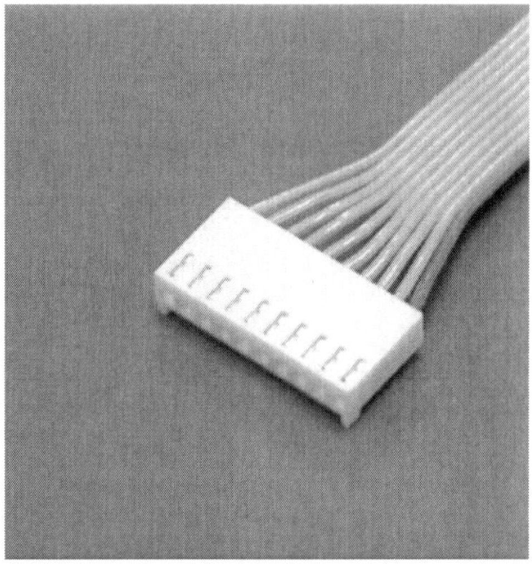

FIGURE 6.4 A ribbon cable for indoor applications.

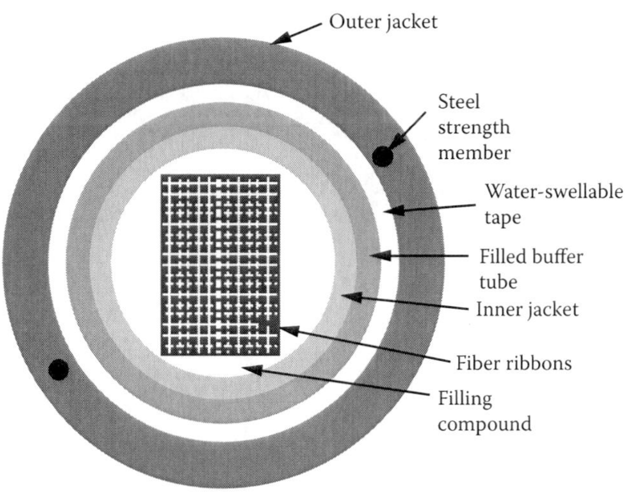

FIGURE 6.5 A cross section of an outdoor ribbon cable.

6.3.1 Indoor Fiber Optic Cable Installation

Generally, indoor cables are placed in conduits or trays. Since standard fiber optic cables are electrically nonconductive, they may be placed in trays alongside high-voltage cables without the special insulation required by copper cabling. Plenum fiber optic cables can be placed without special restrictions in any plenum area within a building.

6.3.2 Cable Installation in Tray and Duct Systems

The primary consideration in selecting a route for fiber optic cable (through trays and ducts) is the avoidance of potential cutting edges and sharp bends. Areas where particular caution must be taken include the corners and the exit slots in the sides of the trays, as shown in Figure 6.6.

If a fiber optic cable is in the same tray or duct as a very large and heavy electrical cable, care must be taken to avoid placing an excessive weight on the fiber optic cable. Figure 6.7 illustrates such a careless case.

Cables in ducts and trays are not subjected to tensile forces. The tensile load must be considered when determining the minimum bend radius at the top of a vertical run. Long vertical runs must be clamped at intermediate points to prevent excessive tensile loading on the fiber cable. The clamping force should not exceed that which

FIGURE 6.6 Fiber optic cables in the tray system.

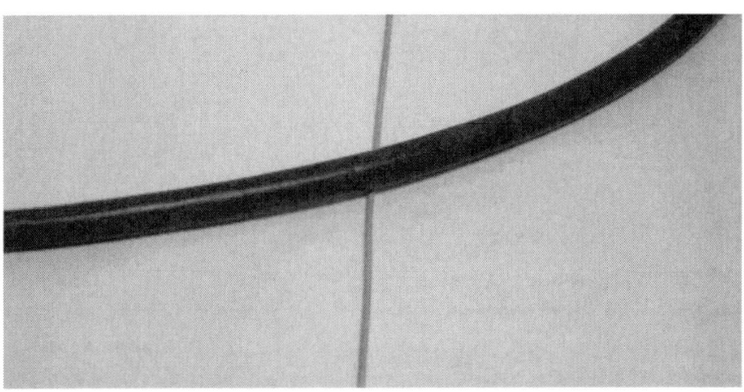

FIGURE 6.7 Heavy power cable crossing over the fiber optic cable.

is necessary to prevent the possibility of slippage. Clamping forces are often determined experimentally, since they are dependent on the type of clamping and jacket materials. The clamping force should be applied over as long a length of the fiber optic cable as is practical. Clamping surfaces should be made of a soft material, such as rubber or plastic. Tensile load during vertical installation is reduced by installing the fiber cable in a top-down manner.

6.3.3 Conduit Installation

Fiber optic cables are pulled through a conduit (wireway or tray) by a wire or synthetic rope attached to the cable. Any pulling force must be applied to the cable strength member and not to the fiber optic cable. In situations where the fiber optic cable does not have connectors, the pull wire should be tied to the Kevlar-strength member, or a pulling grip may be taped to the cable jacket or sheath. When determining the suitability of a conduit for a fiber optic cable, the clearance between the conduit and any other cable is critical (described by a fill factor). Sufficient clearance must be provided that allows the fiber optic cable to be pulled through without excessive friction or binding. If the conduit must make a 90° turn, a fitting such as that shown in Figure 6.8 must be used to allow the cable to be pulled in a straight line while avoiding any sharp bends in the cable.

6.3.4 Pulling Fiber Optic Cable Installation

Fiber optic cables are pulled using many of the same tools and techniques that apply to wire cable installations. Departures from standard methods are required since connectors are usually preinstalled on the fiber optic cable, smaller pulling forces are allowed, and there are minimum bend radius requirements. The pull tape must be attached to the optical cable in such a way that the pulling forces are applied to the outer Kevlar layer. Connectors should be protected to prevent damage. The recommended method for attaching a pulling tape to a simplex cable is the Kellems grip, shown in Figure 6.9. The connector should be wrapped in a thin layer of foam rubber and inserted in a plastic sleeve for protection. The fiber optic cable grip should be

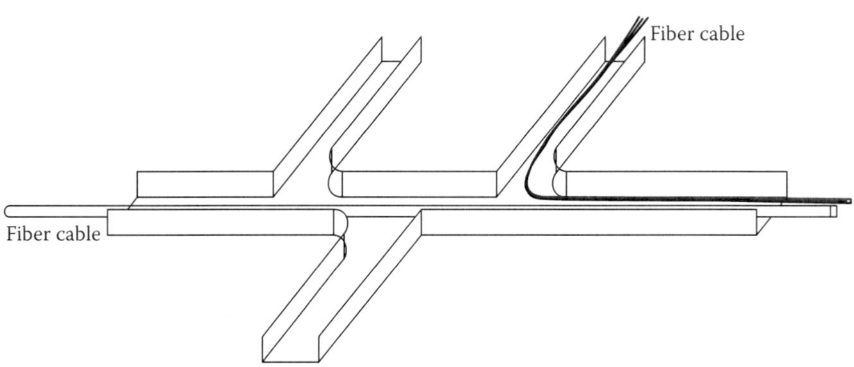

FIGURE 6.8 Wireway or tray with turn fittings.

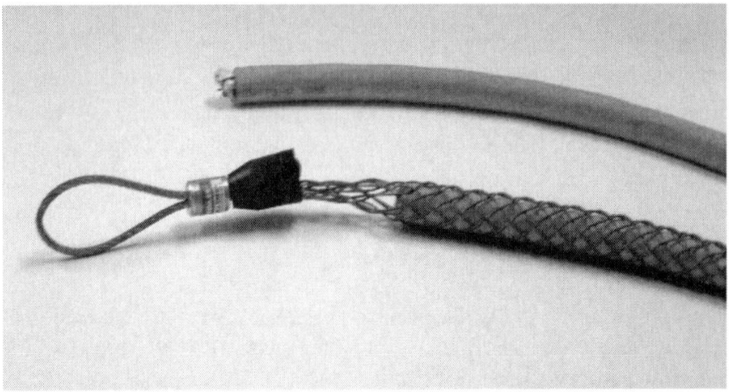

FIGURE 6.9 Kellems grip.

stretched and then wrapped tightly with electrical tape, to provide a firm grip on the fiber optic cable.

The duplex fiber optic cable is supplied with Kevlar-strength members extending beyond the outer jacket to provide a means of attaching the pulling tape. The Kevlar layer is epoxyed to the outer jacket and inner layers, to prevent torsion while the cable is being pulled. The Kevlar is wrapped around the inner jacket in a helical pattern. The free ends of the Kevlar fibers are inserted into a loop at the end of the pulling tape and then epoxyed back to themselves. The connectors are protected by foam rubber and a heat shrink sleeving. The heat shrink sleeving is clamped to the front of the steel ring in the pulling tape, to prevent pushing the connectors along the rest of the fiber optic cable.

During an installation, the pulling force should be constantly monitored by a mechanical gauge. The pull tension can be monitored by a line tensiometer, by breakaway, or by using a dynamometer and pulley arrangement. If a power winch is used to assist the pulling, a power capstan with adjustable slip clutch is recommended. The clutch, set for the maximum loading, will disengage if the set load is reached. The fiber optic cable should be continuously lubricated using gel or powder if necessary.

6.3.5 Fiber Optic Cables Direct Burial Installation

Fiber optic cables can be buried directly in the ground using either plowing or trenching methods, as shown in Figure 6.10. The plowing method uses a cable-laying plow, which opens the ground, lays the cable, and covers the cable in a single operation. In the trench method, a trench is dug with a machine, such as a backhoe, the cable is laid, and the trench is filled. The trench method is more appropriate for short-distance installations. Buried fiber optic cables must be protected against frost, water seepage, burrowing and gnawing animals, and mechanical stresses that could result from earth movements.

Armored fiber optic cables specially designed for direct and indirect burial are available. Cables should be buried so they are below the frost line. Other buried cables should be enclosed in sturdy concrete ducts and polyurethane or PVC pipes,

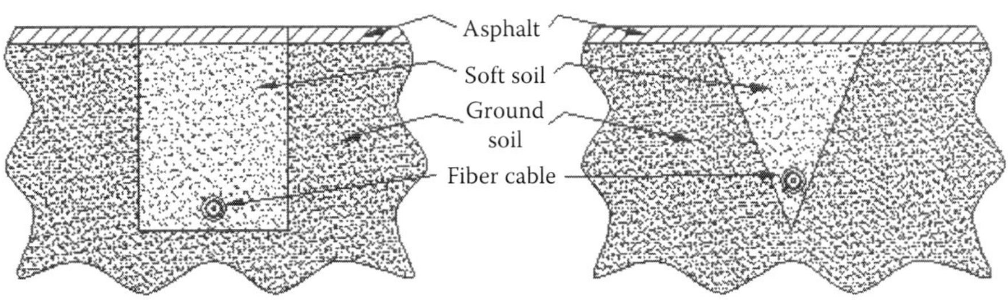

FIGURE 6.10 Fiber optic cable direct burial methods.

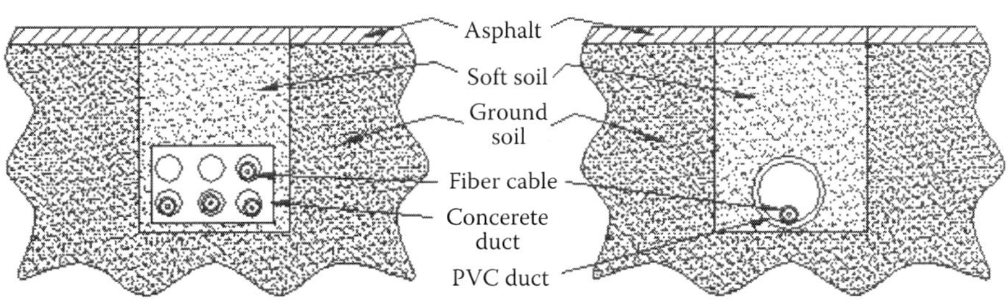

FIGURE 6.11 Fiber optic cable indirect burial methods.

as shown in Figure 6.11. The duct holes and pipes should have an inside diameter several times larger than the outside diameter of the fiber optic cable, to protect against earth movements. An excess length of fiber optic cable in the duct or pipe prevents tensile loads from being placed on the fiber optic cable.

6.3.6 Fiber Optic Cable Aerial Installation

Aerial installation includes stringing fiber optic cables between telephone poles or along power lines, as shown in Figure 6.12. Unlike copper cables, fiber optic cables may be run along power lines with no danger of inductive interference. Aerial fiber optic cables must be able to withstand the forces of high winds, storms, ice loading, and birds. Self-supporting aerial cables can be strung directly from pole to pole.

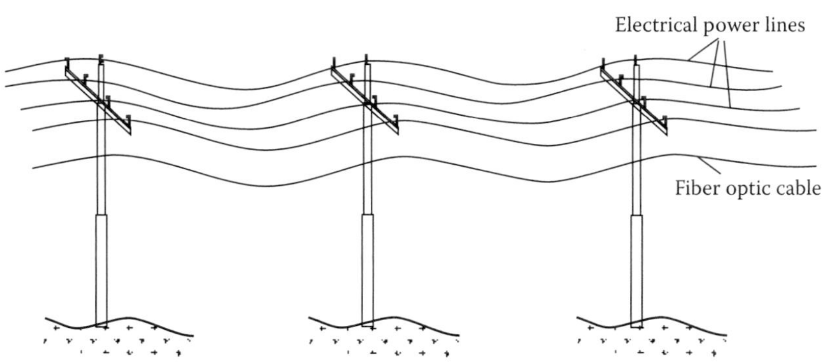

FIGURE 6.12 Fiber optic cable aerial installation.

Other cables must be attached to a high-strength steel wire, which provides additional support. The use of a separate support structure by lashing is the preferred method.

6.3.7 Air-Blown Fiber Cable Installation

There are many ways to upgrade or plan a new network system. Some network architects may choose a new and improved technology. The recent technology of air-blown optical fiber was invented by British Telecom (BT-London) and deployed in the early 1980s. This technology offers installation speed, capacity, and security benefits over other technologies. The air-blown fiber process involves the deployment of tube cable in place of traditional inner ductwork, as shown in Figure 6.13. These tube cables contain several individual tube cells inside a protective outer jacket. The individual tube cells can be in different diameters. Once the tube cable is in place, fiber cable is blown through the tubes to various locations and terminated or interconnected. Fiber in different types (single or multimode) is blown into the network. The air-blown fiber system can work with any panel and connector of traditional fibers. The technology can be used to provide pathways from building to building, city to city, etc. It also allows upgrade, growth, and change of high-speed voice, data, and video in local area networks.

6.3.8 Other Fiber Cable Installation Methods

In 2001, Toronto was the first major city in North America to test drive a system to lay fiber optic cable in sewer pipes deep below the city streets using a remote-controlled robot. With the new technology, the city of Toronto managed to upgrade and extend the communication networks without digging up the streets. The system uses air-blown cables that are installed by a robot. The robot navigates cables and performs all-at-once infrastructure installation. Connections between the points are achieved quickly and efficiently. This technology is planned for use in other cities in Canada.

Cities caught on quickly, and now several cities worldwide enjoy the benefits of having their own fiber networks upgraded. Vienna and Berlin applied this technology to upgrade their network systems.

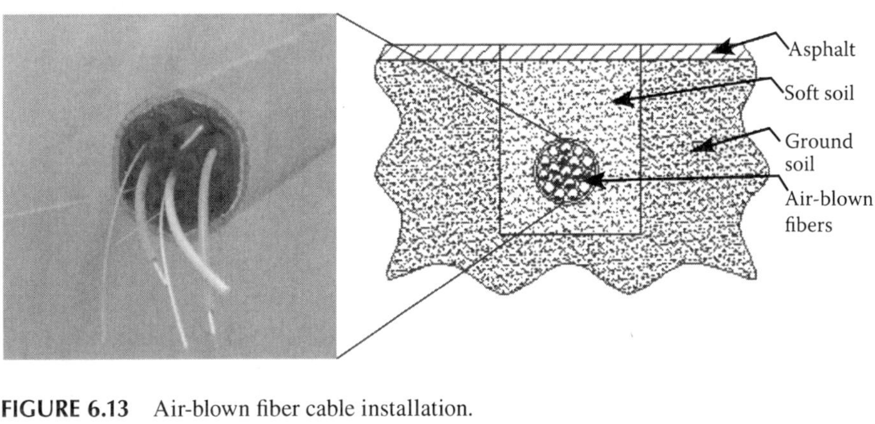

FIGURE 6.13 Air-blown fiber cable installation.

In 2003, the city of San Francisco conducted the 2-mile pilot project. The project enables San Francisco's Department of Telecommunications and Information Services to connect additional city facilities to E-Net. The project uses flexible cable, air-blown tubes, and robots to install the tubes through sewer cleanouts. Each tube houses up to 19 individual fiber tubes. Distribution boxes housing the individual fiber tube connections are strategically placed to avoid unnecessary splice points along the network. This project demonstrates the importance of using a new technology that advances network systems.

6.4 STANDARD HARDWARE FOR FIBER OPTIC CABLES

This section describes some types of common fiber optic hardware that are used in any telecommunication system.

6.4.1 Fiber Splice Closures

There are different types of splice closures for fiber optic cables. A splice closure is a standard piece of hardware in the telecommunication industry for protecting fiber optic cable splices. Some small closure types are for two to eight fiber mechanical splices and are suitable for indoor/outdoor installations. Other closure types include splice trays and large splice closures for outdoor applications. Figure 6.14 shows a rack-mounted splice closure unit.

FIGURE 6.14 A rack-mounted splice closure unit.

6.4.2 Rack with Panels

Consider building an application such as a wiring center that will be used as a central distribution point, as shown in Figure 6.15. Fiber cables can be routed to their final destinations from the rack. An outdoor multifiber cable is brought into the building to the distribution rack, and then simplex or duplex fiber cables route the signals to different locations within the building.

FIGURE 6.15 Rack with panels.

6.4.3 Connector Housings

Connector housings are designed for main cross-connects, interconnects, and intermediate cross-connects, for the local area network and for data center fiber distribution frames in telecommunication systems. Figure 6.15 shows a connector housing unit. They provide easy, open access to connectors for relocations, additions, modifications, and connector cleaning. They can easily be mounted on the racks or cabinets.

6.4.4 Patch Panels

Patch panels provide a convenient way to rearrange fiber cable connections and circuits, as shown in Figure 6.15. A simple patch panel is a metal frame containing bushings in which fiber optic cable connectors plug in on either side. One side of the panel is usually fixed, meaning that the fiber cables are not intended to be disconnected. On the other side of the panel, fiber cables can be connected and disconnected to arrange the circuits as required. Patch panels are widely used in the telecommunication industry to connect circuits to transmission equipment. They can also be used in the telecommunication room of a building to rearrange connections. The splice organizer and the patch panel serve similar functions in distribution. The primary difference is that the organizer is intended for fixed connections, whereas a patch panel is used for flexible connections.

6.4.5 Splice Housings

They provide storage and protection of fiber splices in individually accessible trays. The splice trays are explained in the fiber optic connection chapter. They can easily be mounted on the racks for transition splicing between different cables at building entrance or pigtail splicing.

6.4.6 Wall Outlets

Fiber optic wall outlets serve a similar function to electrical outlets except that they allow connections to fiber cables carrying optical signals, as shown in Figure 6.16. In a building or office wired with fiber optic cables, the outlet serves as a transition point between the horizontal cabling and the equipment. Wall outlets are designed to accommodate the fiber connectors at a 45° mating angle. This design avoids the fiber optic cable bending at the back of the wall outlet. A short simplex or duplex cable, called a jumper or drop cable, typically runs from the wall outlet to the equipment being served.

6.4.7 Fiber Optic Testing Equipment

There are many types of fiber optic testing equipment that offer true multitesting capability. They feature field-installable single-mode and multimode cables, optical time domain reflectometer modules, a visual fault locator, optical return loss, a fiber communicator, an optical network simulator, a built-in power meter, a laser source, etc.

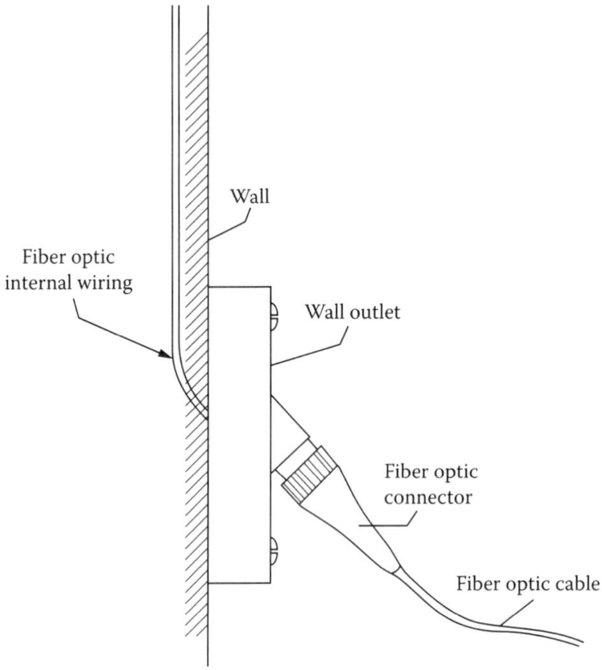

FIGURE 6.16 Wall outlet.

6.5 FIBER OPTIC CABLE TEST REQUIREMENTS

After the fiber cabling system is installed, it should be tested and certified so that it meets performance specifications (TIA/EIA). Testing and certification have different implications when installing fiber optic cables. Testing implies that certain values are measured. Certification, on the other hand, compares these measured values to standards or derived values to determine that the values are within specified limits. Testing is a quantitative procedure, while certification is both quantitative and qualitative.

Attenuation tests at different operating wavelengths should be performed on all fiber optic cables. The test results should be documented, comply with the standards, and include the following information:

- Fiber optic cable ID
- Fiber optic cable types
- Test equipment
- Testing and verification techniques
- Attenuation values
- Wavelengths
- Connector types
- Connector names
- Splice types and locations
- Reference setting at first wavelength

- Reference setting at second wavelength
- Building number and location
- Mapping and documentation requirements
- Relevant additional data
- Relevant additional comments
- Any special customer requirements
- Test results
- Date of test performed
- Contractors' names
- Technicians' names and signatures
- Any other requirements that are specified by the customer

6.6 HOW TO INSTALL A FIBER OPTIC CABLE

There are many types and applications of fiber optic cable. The practical importance of the mechanical properties of fiber optic cable is demonstrated. The different types of common fiber optic hardware, such as closures, organizers, rack boxes, distribution panels, and wall outlets, play an important role in telecommunication systems. The student will also study and practice the parts of the telecommunication system in the building. The student will conduct the correct manner of installing and testing a fiber optic cable in a networking system.

6.7 TECHNIQUE AND APPARATUS

Figure 6.17 shows the tools that are required for this job. **Note:** Tools and parts are listed in items 9 to 12 in case they are needed to build a connector or splice as an adapter between the fiber optic cable and the testing devices.

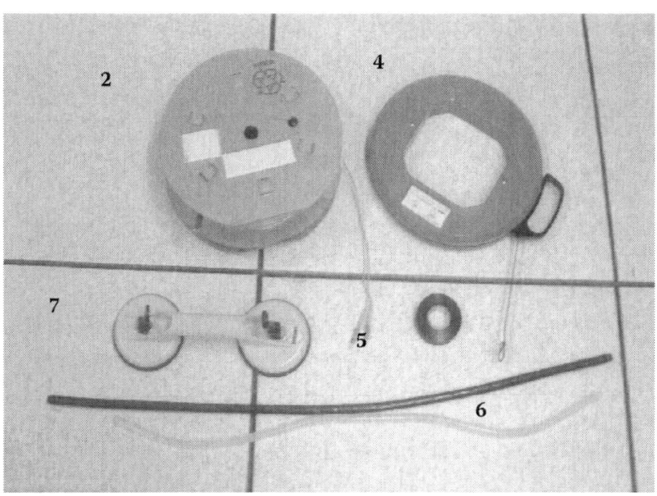

FIGURE 6.17 Fiber cable installation tools.

1. Fiber optic cable
2. Fish wire
3. Plastic tape/isolation tape
4. Protective plastic spiral tube
5. Vacuum handle for removing floor tiles
6. Optical continuity checker
7. Optical power meter and sensor
8. Fiber cable measurement set
9. Connector assembly kit
10. Mechanical optical splice kit
11. Fusion splice

6.8 FIBER OPTIC CABLE INSTALLATION PROCEDURE

Choose the area where the cable should be installed. Try to choose two types of cable installation procedures between two connections, such as an overhead cable tray system and an installation under raised floor tiles. The following steps will guide you throughout this experiment:

1. Locate the area and plan your fiber optic cable installation in the telecommunications lab.
2. Determine the route that will be taken by the fiber optic cable.
3. Ensure proper coiling of the fiber optic cable. Arrange all of your cable in the correct entry position using a spiral rubber tube to protect the fiber optic cable from any sharp edges.
4. Remove every second tile from the floor, starting at the termination/distribution box along the fiber optic cable raceways.
5. Properly attach the fiber optic cable to the fish wire after passing it through the protective jacket in case of excessive bending.
6. Pull slowly and have someone assist with feeding the fiber optic cable from the other end.
7. Proceed to pull the fiber optic cable through the overhead cable tray using the fish wire.
8. Keep the fiber optic cable neatly bundled, laying it on the floor away from posts and other obstacles.
9. Carefully pull the fiber optic cable through the tray and position it in the center of the cable tray. Trim the excess fiber optic cable, leaving sufficient service length of the cable for repair or maintenance.
10. Replace the tiles back into position, tidy up the site, and store the tools.
11. Terminate each cable end and make the required end connections.
12. Carry out the fiber optic cable continuity test.
13. Conduct cable end connection tests.
14. Measure the fiber optic cable power loss.

7 Manufacturing of Passive Fiber Optic Devices

7.1 INTRODUCTION

A variety of passive devices are used in communication systems to provide specific functions. Passive devices work without using an external power supply, while active devices need external power to work. Some common passive fiber optic devices are presented in detail in this chapter. Passive devices split, redirect, or combine light waves.

The most simple passive fiber optic devices are the couplers. Couplers couple multiple input light waves to multiple outputs. Normally, the couplers split input signals into two or more outputs, or combine two or more inputs into one output. Couplers can have more than two inputs/outputs. Couplers work as power splitters, power taps, and wavelength selectors.

Wavelength division multiplexers and demultiplexers, filters, circulators, and isolators are also called passive devices. This chapter presents one section that enables the student to practice designing, manufacturing, and testing a 50/50 Y-coupler in the lab. The student can plan the manufacturing process and determine the process and handling times that are required to build one Y-coupler. The student will manufacture and test a Y-coupler in the lab.

7.2 2 × 2 COUPLERS

In an optical network, there are many situations where it is necessary to combine or split light signals. Directional couplers form the basis of many data distribution networks. Figure 7.1 shows a four-port directional coupler with two inputs and two outputs. The arrows indicate the directions of power flow through the coupler. Power P_1 is incident on the input port, port 1, of the coupler; the signal is divided between output ports, ports 2 and 3, based on a set splitting ratio. Ideally, no power will reach port 4, called the isolated port. By convention, power P_2 emerging from port 2 is equal to or greater than power P_3 emerging from port 3, depending on the designed splitting ratio of the coupler.

The main parameters of couplers are optical power losses. There are several types of coupler losses, given in unit of decibels (dB):

1. Throughput loss (L_{THP}) specifies the transmission loss between input power P_1 at port 1 and transmission power P_2 at port 2. This loss is calculated by

$$L_{THP} = -10 \log_{10}\left(\frac{P_2}{P_1}\right) \tag{7.1}$$

2. Tap loss (L_{TAP}) specifies the transmission loss between input power P_1 at port 1 and tap power P_3 at port 3. This loss is calculated by

$$L_{TAP} = -10 \log_{10}\left(\frac{P_3}{P_1}\right) \tag{7.2}$$

3. Directional loss (L_D) specifies the loss between input power P_1 at port 1 and isolated power P_4 at port 4. Ideally, the isolated power loss is zero. This loss is calculated by

$$L_D = -10 \log_{10}\left(\frac{P_4}{P_1}\right) \tag{7.3}$$

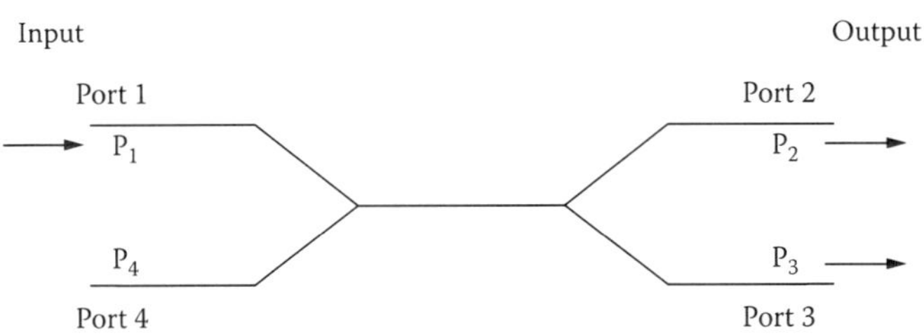

FIGURE 7.1 A four-port directional coupler.

TABLE 7.1
Characteristics of Several Common Couplers

Coupler Description	Splitting Ratio	L_{THP} (dB)	L^{TAP} (dB)
3 dB	1:1	3	3
6 dB	3:1	1.25	6
10 dB	9:1	0.46	10
12 dB	15:1	0.28	12

4. Excess loss (L_E) specifies the power lost within the coupler. It includes radiation, scattering, absorption, and coupling to the isolated port. This loss is calculated by

$$L_E = -10\ \log_{10}\left(\frac{P_2 + P_3}{P_1}\right) \tag{7.4}$$

Table 7.1 lists splitting ratio values, throughput loss, and tap loss for several common couplers.

7.3 3 DB COUPLERS

A simple four-port coupler is often called a 3 dB coupler if the input light splits into two equal portions at the output ports. The signal is split in half (3 dB = half). The 3 dB comes from the power loss formula: [$--10 \times \log_{10} (P_2/P_1) = -10 \times \log_{10} (0.5/1.0) = -10 \times (-0.3) = 3$ dB]. Half of the light entering at port 1 will exit at port 2 and half at port 3, as shown in Figure 7.2.

It is often useful to cascade many 3 dB couplers, as shown in Figure 7.3. The configuration shown is called a splitter. This splitter divides a single input into four equal outputs and is denoted as a 1 × 4 coupler. As might be expected, if the device is perfect, each output port will contain one-fourth of the input power.

Figure 7.4 shows how a 1 × 8 coupler can be constructed by cascading several 3 dB couplers in a tree configuration. The signal input power will divide into eight equal outputs. Each output port will contain one-eighth of the input power.

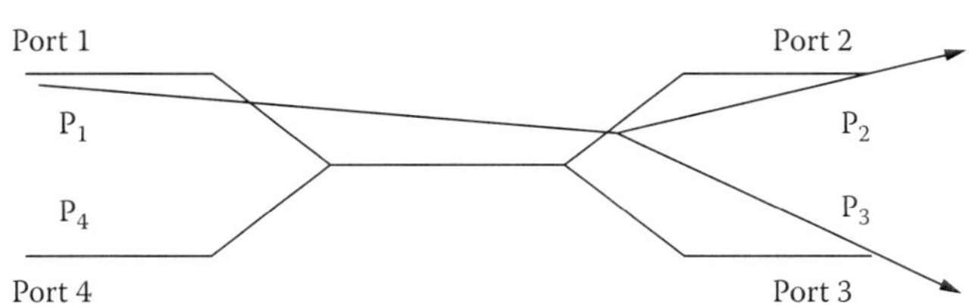

FIGURE 7.2 3 dB coupler.

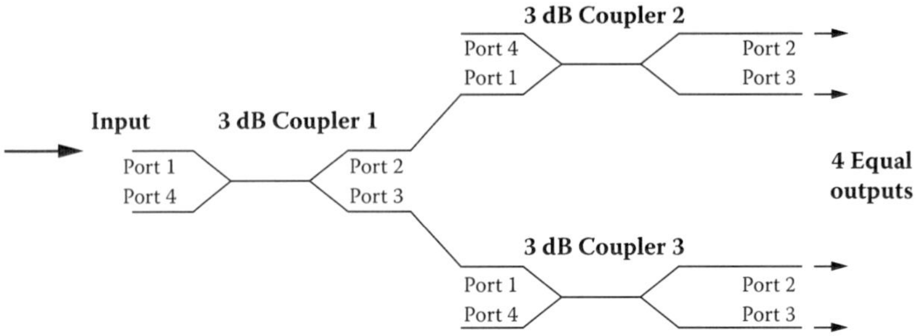

FIGURE 7.3 Cascaded 3 dB couplers to produce a 1 × 4 coupler.

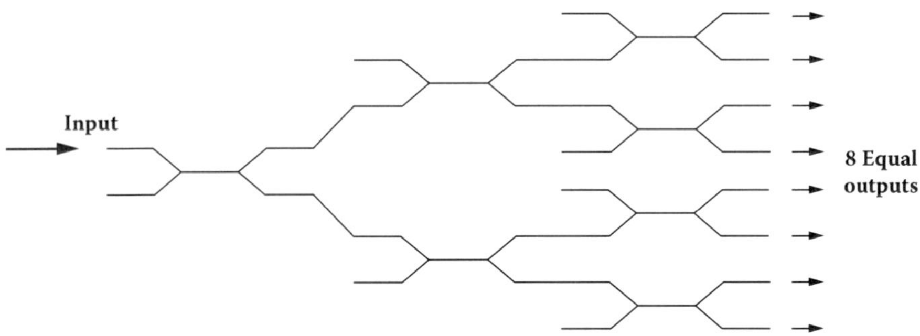

FIGURE 7.4 Cascaded 3 dB couplers to produce a 1 × 8 coupler.

7.4 Y-COUPLERS

Y-couplers or splitters, sometimes called 3 dB couplers, split the light equally. Y-couplers are 3 dB couplers in which port 4 is not used. In the Y-coupler (splitter), as shown in Figure 7.5, the light entering port 1 will be split equally between ports 2 and 3 with almost no loss. They are extremely efficient at splitting light with little loss. Y-couplers are difficult to construct in fiber optics, but they are easy to

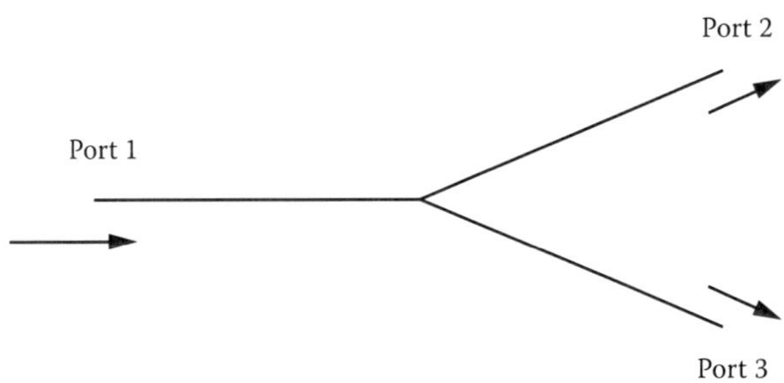

FIGURE 7.5 Y-coupler (splitter).

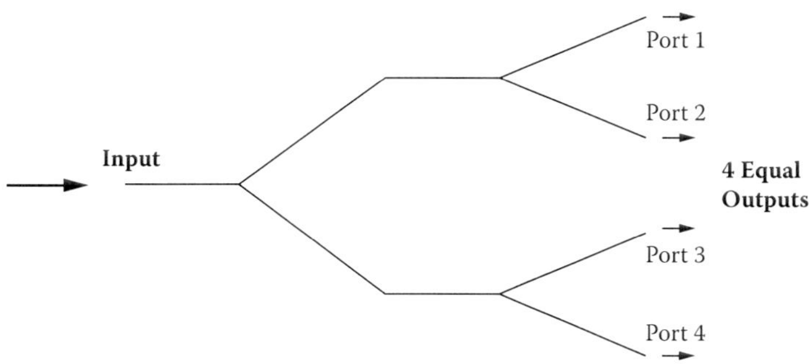

FIGURE 7.6 Cascaded 1 × 2 Y-couplers to produce a 1× 4 coupler.

use in planar waveguide systems. The power loss in the Y-coupler system can be calculated by

$$Loss\,(dB) = -10\,\log_{10}\left(\frac{P_{\text{out}}}{P_{\text{in}}}\right) \tag{7.5}$$

Y-couplers are very seldom built as separate planar devices; instead, they are manufactured on the same substrate as other devices. Connecting Y-couplers to a fiber optic cable is expensive, and significant loss is experienced in the connections. However, Y-couplers of this kind are used extensively in complex planar devices.

It is often useful to cascade Y-couplers, as shown in Figure 7.6. The splitter configuration shown divides a single input into four equal outputs. If the device is perfect, each output port will contain one-fourth of the input power. Figure 7.6 shows how a 1 × 4 Y-coupler can be constructed by cascading three 1 × 2 Y-couplers in a tree arrangement.

Figure 7.7 shows how a 1 × 8 Y-coupler can be constructed by cascading seven Y-couplers in a tree configuration. The power of the input signal will divide into eight equal outputs. Each output port will contain one-eighth of the input power.

7.5 STAR COUPLERS

A star coupler is simply a multiple-output coupler in which each input signal is made available on every output fiber cable. There are two star coupler designs, as shown in Figure 7.8. Figure 7.8(a) shows an 8 × 8 coupler. This coupler distributes the power from any input port to all of the output ports, splitting equally among the output ports. This type of coupler is called a transmission star coupler. Figure 7.8(b) shows a reflection star coupler; any input is split equally among all fiber cables. Star couplers are typically used in local area networks (LANs) and metropolitan area networks (MANs).

7.6 COUPLER CONSTRUCTION

In practice, there are many techniques to construct couplers. Figure 7.9 shows the most common designs of manufactured couplers.

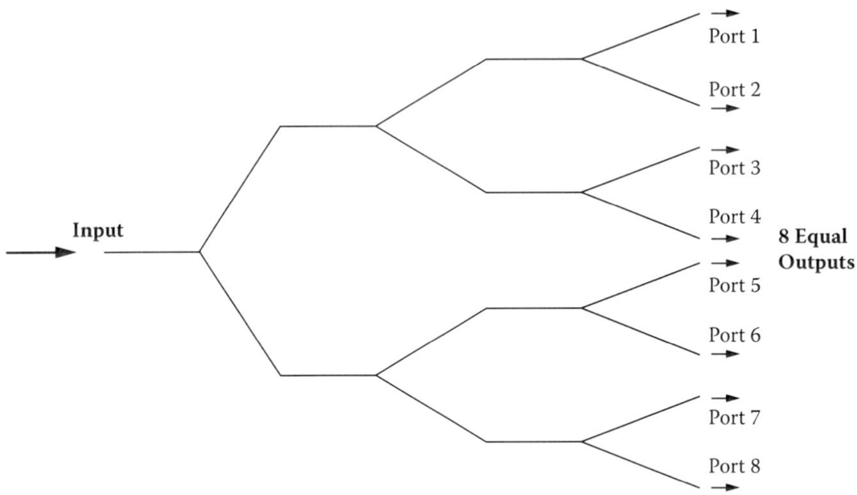

FIGURE 7.7 Cascaded Y-couplers to produce a 1 × 8 Y-coupler.

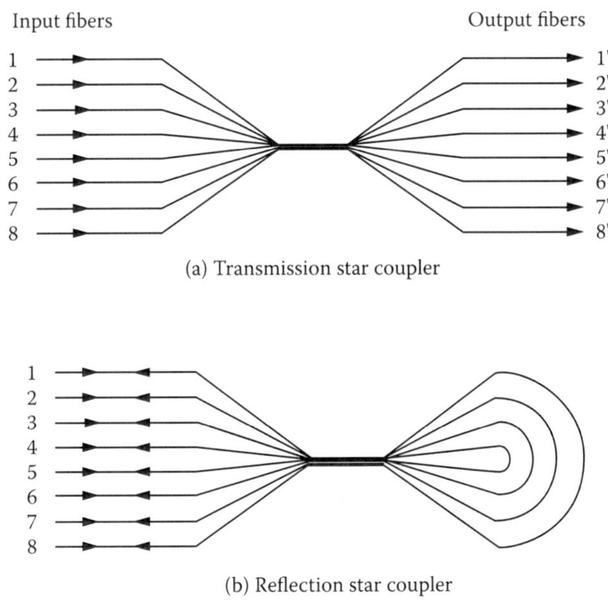

FIGURE 7.8 Star couplers.

7.6.1 Fused Taper Couplers

Fused taper couplers consist of two regular single-mode fiber optic cables, which are in contact with one another, as shown in Figure 7.9. In this design, two fiber cables are heated, and then tension is applied to the ends of the fiber cables. Both the claddings and the cores of the fiber cables are drawn, and thus become thinner. During

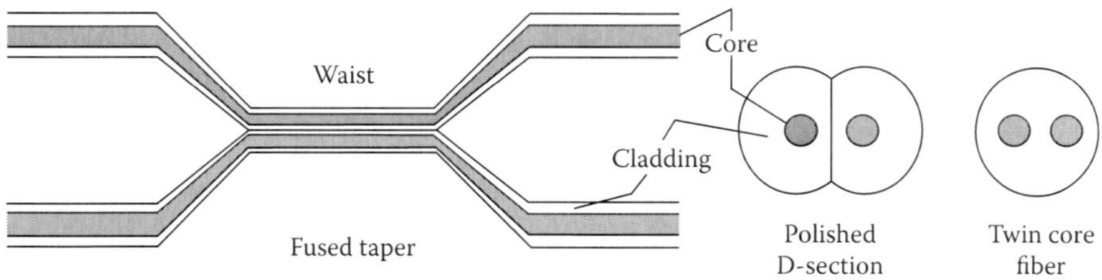

FIGURE 7.9 Some coupler configurations.

this process, the fiber cables fuse together, forming the waist. Sometimes the fiber cables are twisted tightly together before heating and stretching. Fused taper couplers are very common commercial devices.

7.6.2 Polishing D-Section Couplers

The design of the polishing D-section couplers involves embedding fiber cables into a solid material such as plastic. The cladding surface is polished along the length of the fiber cables until one side of the cladding is flat and removed to within 4 microns of the core. A D-shaped fiber cable section results, as shown in Figure 7.9. The plastic is then dissolved away. The two sections are joined along their flat surfaces using index matching epoxy resin. Sometimes, the cladding on the fiber cables is thinned out by etching with hydrofluoric acid to reduce the amount of polishing. This is a relatively precise technique. However, it is more expensive to make than the fused taper method.

7.6.3 Twin Core Fiber Couplers

The design of the twin core fiber couplers uses twin cores having a very small separation, as shown in Figure 7.9. The manufacturing cost of the twin core fiber cable is very low compared to the cost of other designs. However, it is more difficult to connect twin core couplers to regular fiber cables than to other coupler designs.

8 Manufacturing of Active Fiber Optic Devices

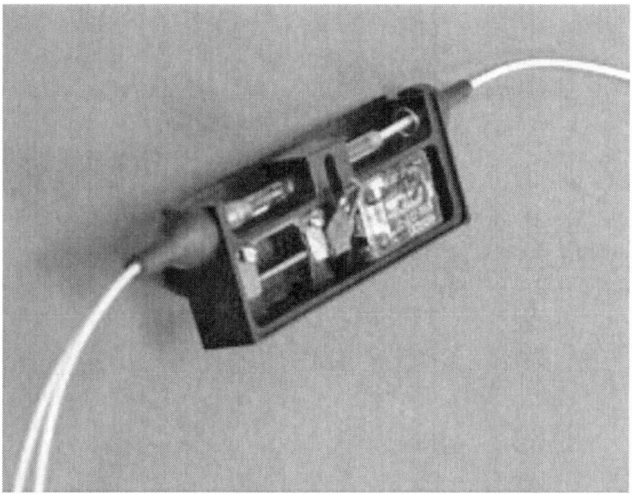

8.1 INTRODUCTION

Many optical communication networks integrate optical switches into their design. Optomechanical switches redirect optical signals from one port to another by moving a fiber tube assembly or an optical component, such as a mirror or prism. There are many different types of optical switches incorporated into networks. In practice, most optical switches are still operated mechanically and controlled by an electronic control circuit. Speed is a crucial parameter in network applications, since a high-speed data transmission of tenths of milliseconds is required. In the near future, dynamic optical routing will require much faster switching speeds.

More technology exists for optical switches than for any other functional component within the optical network. Researchers are developing optical switches to increase the number of outputs and to reduce size, cost, and switching time. Presently, optical switches include many types, for example: optomechanical switches, thermo-optic switches, electro-optic switches, microelectromechanical switches (MEMS), and micro-optomechanical switches (MOMS). New types of optical switches are in the research and development stages.

This chapter illustrates a few switch designs, which are manufactured for use in optical communication systems and other applications. Optomechanical and electromechanical switches are the oldest type of optical switch and the most widely deployed at this time. These devices achieve switching by moving fiber or other optical components by means of stepper motors or relay arms. This causes them to have a relatively slow switching time. However, they can achieve excellent reliability, and still offer low insertion and cross talk losses.

This chapter presents four cases in building optomechanical switches using a movable mirror or prism to switch between the input and output ports.

8.2 OPTOMECHANICAL SWITCHES

Figure 8.1 illustrates common switch configurations. The input signal comes through the input fiber cable on the left side of the switch. A mechanical slider moves that fiber up and down, latching into one of the two output fiber cables on the right side of the switch. In OFF/ON positions, the switch directs light from the input fiber into one of the two outputs. This arrangement is called 1 × 2 switch configuration. As input at port 1, the signal can be switched to either port 2 or port 3.

For the following definitions, assume the switch is configured to couple to port 2. The insertion loss L_{IL} (in decibels) is defined by Equation 8.1. Insertion loss depends on fiber cable alignment at the input and output ports. A low insertion loss value can be obtained on switches with good mechanical alignment. A good switch provides similar values of insertion loss for all switch positions.

$$L_{IL} = -10 \times \log_{10} \frac{P_2}{P_1} \tag{8.1}$$

where P_1 is the power going into port 1 and P_2 is the power exiting from port 2.

Cross talk loss L_{CT} is one of the important losses, which should be considered in optomechanical switches. Cross talk loss is a measure of how well the uncoupled port is isolated. The cross talk loss L_{CT} (in decibels) is defined by Equation 8.2. Cross talk loss values depend on the particular design of the switch.

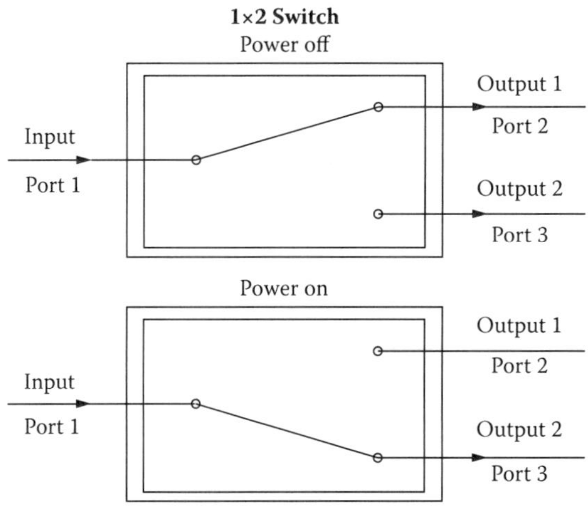

FIGURE 8.1 A typical 1 × 2 switch configuration.

$$L_{CT} = -10 \times \log_{10} \frac{P_3}{P_1} \tag{8.2}$$

where P_1 is the power going into port 1 and P_3 is the power exiting from port 3.

There are other important optical parameters that need to be specified for each switch type. These parameters include polarization-dependent loss (PDL), return loss (RL), and the Etalon effect. The PDL is defined as the maximum difference in insertion loss between any two polarization states. It is caused by mechanical stress and temperature variation on optical components or fiber cables. This causes changes in the birefringence and a gradient of index of refraction (n) of the optical material. The RL is defined as the light reflected back into the input path. It is caused by scattering and reflection from optical surfaces like mirrors, lenses, and connectors or from defects, such as cracks and scratches. The backreflection is equal to the return loss with a negative quantity. The Elaton effect is defined as light resonance (ripple) at a certain wavelength. It is caused by reflection of light from parallel optical surfaces and interference between the signals. All the above losses are measured in decibels (dB). Special optical parameters can be specified by the customers.

Another important parameter of the optical switches is the repeatability—achieving the same insertion loss each time the switch is returned to the same position. Switching speed is also another important specification of a switch. The switching speed is defined as how fast the switch can change the signal from one port to the other. It is an important factor in some switch applications in communication systems.

Figure 8.2 shows a schematic diagram of a mechanical switch configuration with two inputs and two outputs. The inputs are located on the one side and the outputs on the other side of the switch. This configuration is called a 2 × 2 switch. The signal

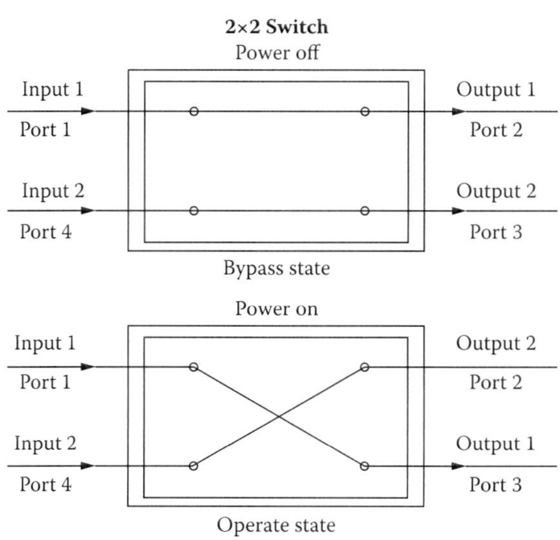

FIGURE 8.2 A typical 2 × 2 switch configuration.

enters ports 1 and 4, and exits from ports 2 and 3, respectively. This case is called the bypass state in the OFF position. When the latching mechanism changes position between ports 2 and 3, signal enters port 1 and exits port 3, and from port 4 to port 2. This case is called the operate state, in the ON position.

Optomechanical switches collimate the optical beam from each input and output fiber and move these collimated beams around inside the switch. This creates low optical loss, and allows distance between the input and the output fiber. These switches have more bulky components than newer alternatives, such as the micro--optomechanical switches.

Figure 8.3 shows a schematic diagram of a two-position switch. The switch consists of a sliding prism and quarter-pitch graded-index lenses (GRIN lens) at the input and output ports. The components are assembled in a packaging base and sealed with a lid. Each GRIN lens is connected to the fiber tube assembly using an epoxy. Figure 8.3 illustrates the OFF/ON positions of a 1 × 2 switch. As explained above, the GRIN lens collimates the divergence beam exiting from the input fiber. The right angle prism deflects the light by total internal reflection at its two slanting

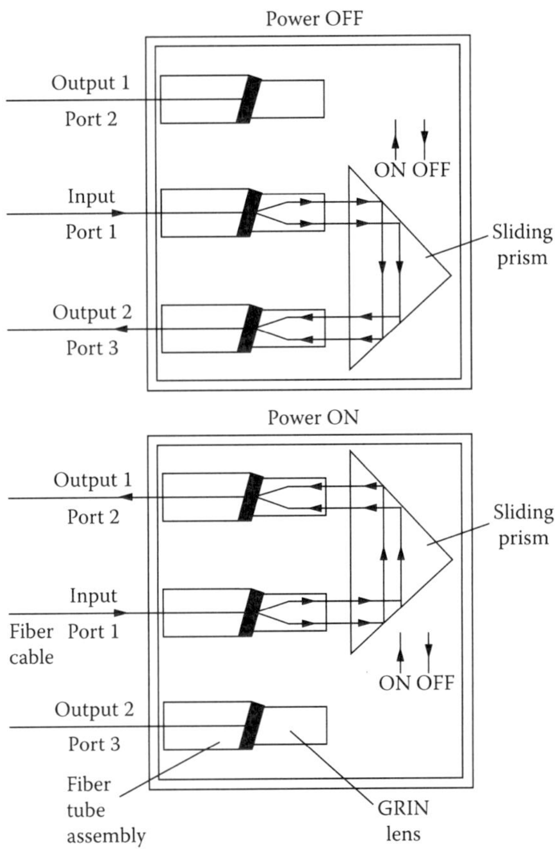

FIGURE 8.3 A 1 × 2 optomechanical switch.

surfaces. The GRIN lens refocuses the collimated beam onto a fiber cable at one of the output ports. To direct the signal from port 1 to port 3, the prism slides to a new position, as shown in Figure 8.3 in the OFF position. Figure 8.3 also shows the signal directed between ports 1 and 2, in the ON position, when the prism changes position.

Optomechanical switches drive optical fiber networks mechanically. They can switch between light paths at high speed and with low insertion loss. They are widely used in rapidly developing areas of the fiber optic field, such as optical cross-connection and wavelength multiplexing. Figure 8.4 shows the design of a 2 × 4 optomechanical switch. The switch uses an electromagnetic actuator with

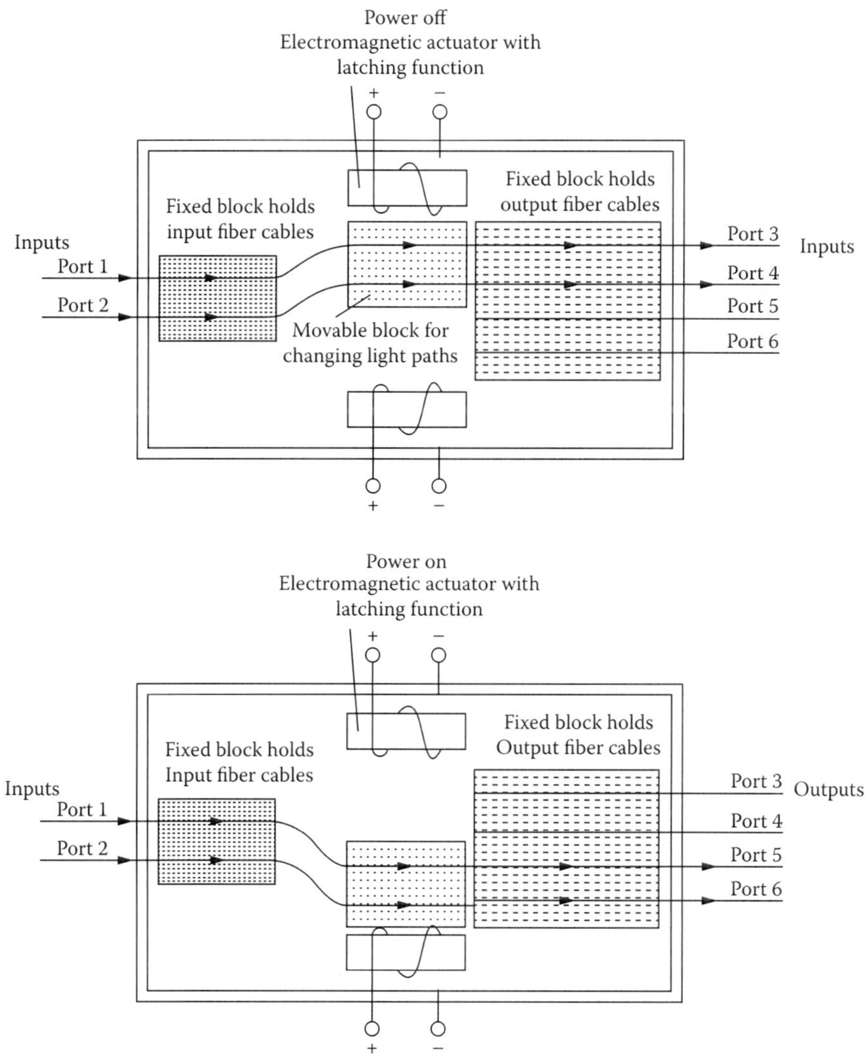

FIGURE 8.4 A 2 × 4 optomechanical switch.

a latching function to drive a movable block to change the light path between the ports. Figure 8.4 shows the switch in the OFF position. The light passes through from ports 1 and 2 to ports 3 and 4, respectively. When the power is turned ON, an electromagnetic actuator (with latching function) drives a movable block to change light path from ports 1 and 2 to ports 5 and 6, respectively, as shown in Figure 8.4. The optical and mechanical components of a switch are assembled in a packaging box with minimal alignment work. There are three configurations of this switch: 1 × 2, 2 × 2, and 2 × 4.

A practical electromagnetic bypass switch is illustrated in Figure 8.5. The switch contains a quarter-pitch GRIN lens connected to fiber tube assembly at the input and output ports, a relay, and an iron bar with mirror end faces. The components are assembled in a packaging base, which is sealed with a lid. When the power is turned OFF, a spring pulls the iron bar out of the signal path, returning the switch to the bypass condition. This is called the bypass state. In the bypass state, the signal passes directly from ports 1 and 4. When the power is ON, the electromagnet is activated and the iron bar is raised. This is called the branch state. In the branch state, mirrors direct the signal between ports 1 and 2, and between ports 3 and 4.

Another type of bypass switch is also used in communication networks. Figure 8.6 illustrates the function of this type of bypass switch. When the power is in the OFF position, the input signal comes through the input fiber cable on port 1 on the left side and leads to output port 4 on the right side of the switch. This is called the bypass state. When the power is ON, a mechanical slider moves two fiber connections to the UP position, latching into two output fiber cables at ports 2 and 3 on the side of the switch. In this position, the input signals from ports 1 and 4 are launched into ports 2 and 3, respectively. This position is called branch state. As input is at port 1, the signal can be directed to either port 4 or port 2. Also, an input signal at port 4 can be directed to output port 3.

This section presents new switch designs of a 1 × 8 latching switch configuration using prisms. These switches are commercially available on the market. There are two types of models: the linear and the triangular models. The linear model directs the signal from the input to the outputs by arranging the prisms linearly, as shown in Figures 8.7 and 8.8. The triangular model directs the input to the outputs by arranging the prisms triangularly, as shown in Figures 8.9 and 8.10. These models have come into wide use because they are simple, offer eight outputs, and are cost-effective. They are also used in backup systems to reroute signals around broken fiber optic cable and in fiber optical instruments.

Figure 8.7 illustrates a schematic diagram for one configuration of a linear model of a 1 × 8 latching switch. The common element of this type of optomechanical switch is that its operation involves the mechanical sliding motion of prisms in OFF/ON positions to direct the signal from one port to another. Figure 8.7 shows the input located on the one side and the outputs on the other side of the switch. This configuration is called a 1 × 8 switch in the linear model. Light enters port 1 and exits from port 2 when sliding prism 1 is in the OFF position. When the latching mechanism places the sliding prisms 1 and 2 in position, the light enters port 1 and exits from port 3, as shown in Figure 8.8. Similarly, light enters port 1 and exits from port 4, when sliding prism 2 is in the OFF position and sliding prism 3 is in the

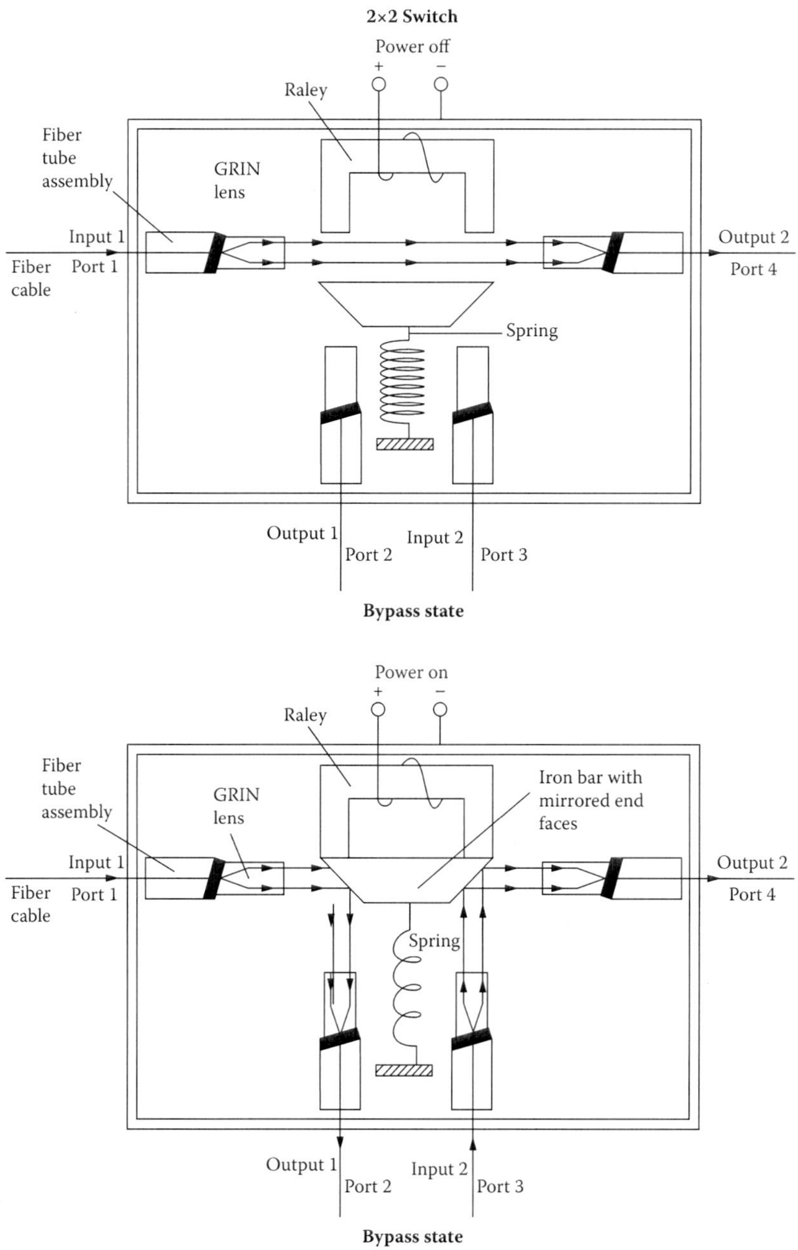

FIGURE 8.5 An electromagnetic bypass switch.

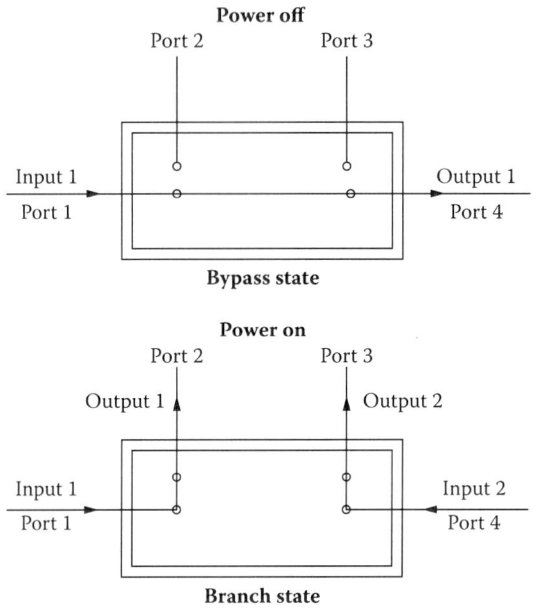

FIGURE 8.6 A bypass switch.

ON position. This switch configuration is more complicated than the second switch configuration because the input is located on one side and the outputs on the other side of the switch. This configuration includes a complex mechanism, controls, seven sliding prisms, and one fixed prism.

Figure 8.9 illustrates a schematic diagram of the second configuration of the linear model of a 1 × 8 latching switch. This configuration is different because the input and outputs are located on the same side of the switch, as shown in Figure 8.9. Prisms are also used in the operation in this type of switch configuration. Light enters port 1 and exits from port 2 when fixed prism 1 and sliding prism 2 are in position. When the latching mechanism places sliding prism 2 in the OFF position and sliding prism 3 in the ON position, the light enters port 1 and exits port 3. Similarly, light enters port 1 and exits from port 4 when sliding prism 3 is in OFF position and sliding prism 4 is in the ON position. The same procedure is used for the signal exiting from other ports. This switch configuration is simpler than the first configuration because the input and outputs are located on the one side of the switch. This configuration includes a less complex mechanism, controls, seven sliding prisms, and two fixed prisms.

Figure 8.10 illustrates a schematic diagram of a configuration of the triangular model of a 1 × 8 latching switch. The common element of this type of optomechanical switch is that the operation involves mechanical sliding motion of parallelogram prisms in OFF/ON positions to direct the signal from one port to another. Figure 8.10 shows the input located on one side and the outputs on the other side of the switch. This configuration is called a 1 × 8 switch in the triangular model. Light enters port 1 and exits from one of the outputs. When the latching mechanism places

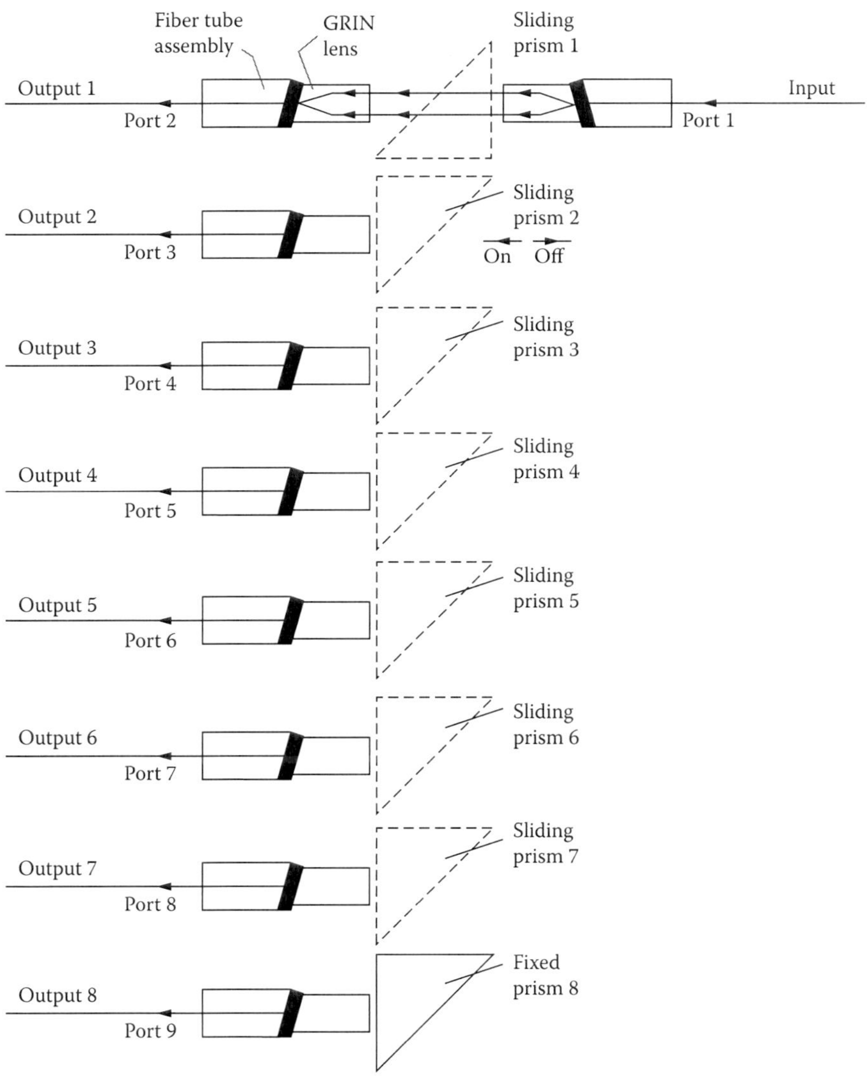

FIGURE 8.7 One configuration of the linear model of a 1 × 8 latching switch.

sliding parallelogram prism 1 in the OFF position and sliding prism 2 into position, the light enters port 1 and exits port 4, as shown in Figure 8.10. Similarly, light enters port 1 and exits from port 3 when sliding prisms 2 and 3 are in the ON position. Figure 8.11 shows light entering port 1 and exiting from port 5, when sliding prism 1 is in the OFF position. This switch configuration is more complicated than the linear model because the input is located on the one side, and the outputs on the other side of the switch. Seven sliding parallelogram prisms with additional mechanisms and controls form this configuration. Both models have challenges achieving precise alignments and low losses during the manufacturing process.

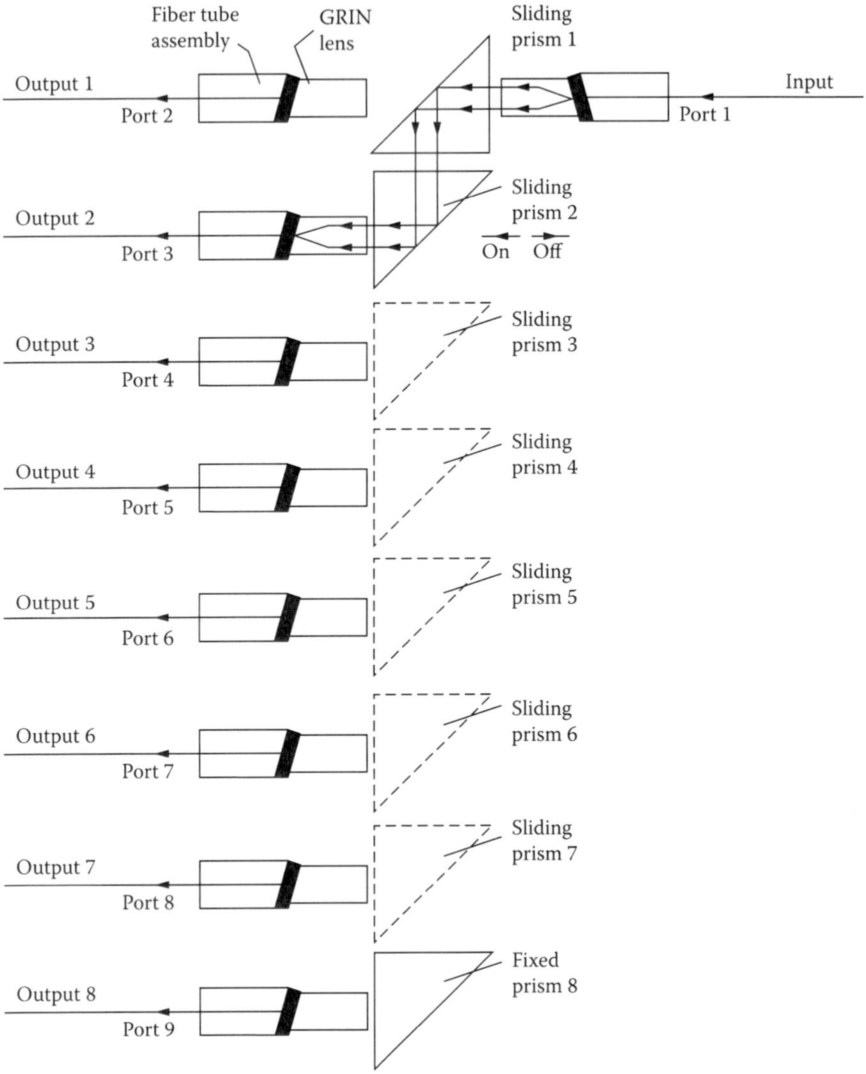

FIGURE 8.8 Signal from port 1 to port 3 in the linear model of a 1 × 8 latching switch.

Many other modern optomechanical switches are used in telecommunication network management, monitoring, restoration, and protection. They have excellent optical performance and high reliability necessary for network applications. They feature low insertion loss, high return loss, channel isolation, excellent repeatability, and fast switching speeds. The switches are available in single mode and multimode, and cover wide wavelength ranges. They are available in 1 × 1, 1 × 2, and 2 × 2 configurations. The switching mechanism is able to latch and remain in its selected state following a loss of power. The switch consists of a quarter-pitch graded-index lens (GRIN lens) glued to a double-bore fiber tube assembly, relay, and mirror mounted on a shaft. The components are assembled inside a packaging base box and covered

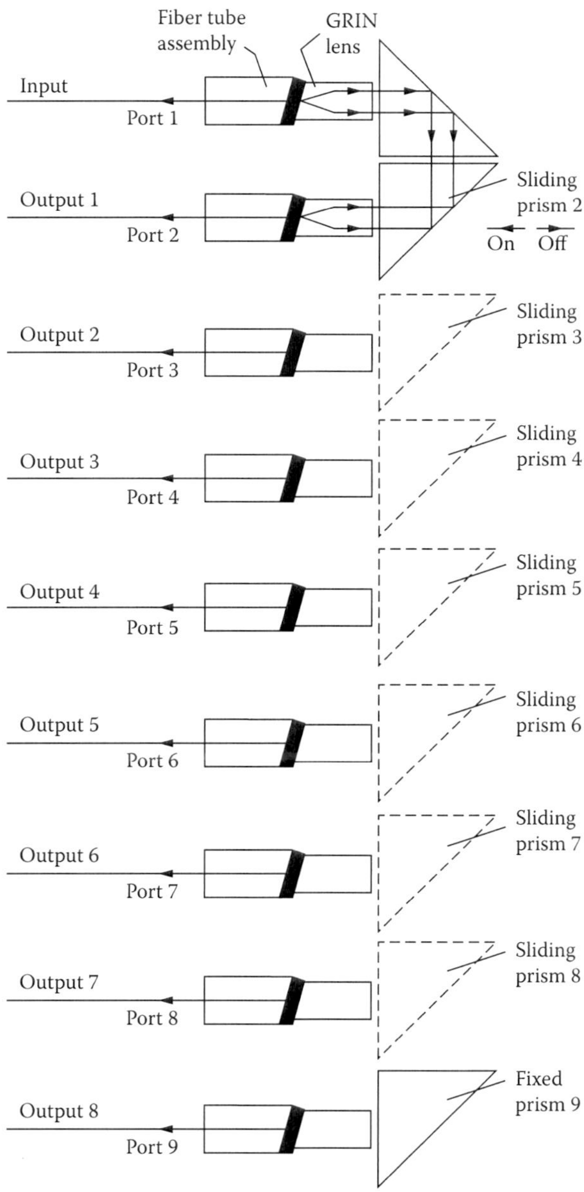

FIGURE 8.9 Second configuration of the linear model of a 1 × 8 latching switch.

by a lid. Figure 8.12 illustrates a schematic diagram of a 2 × 2 switch. This figure illustrates the switch in the OFF/ON positions. When the power is OFF, light transmits from port 1 to port 2 and port 4 to port 3. This configuration is called the transmission state. When the power is ON, the mirror is in position, and the light is reflected by the mirror. Light exits port 1 and reflects to port 4, and similarly port 2 reflects to port 3. This is called the reflection state.

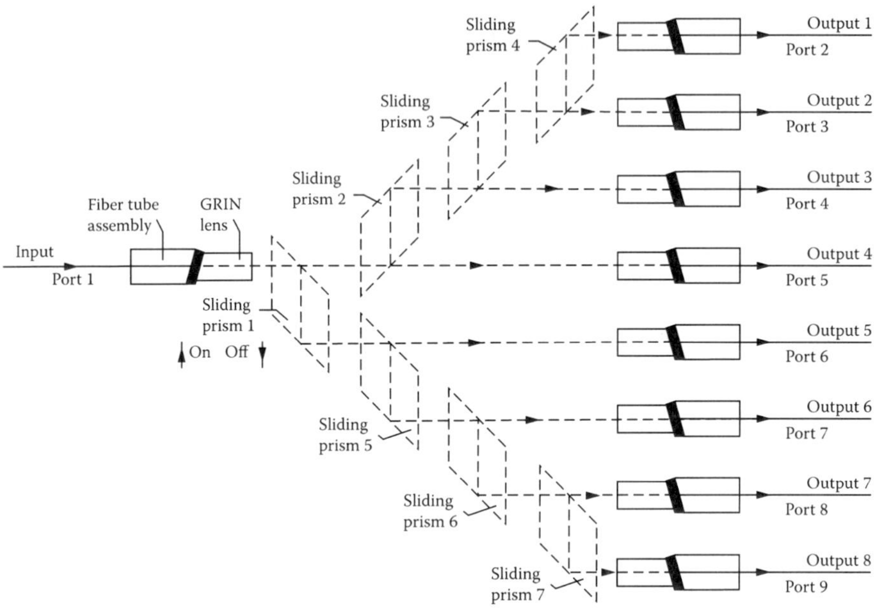

FIGURE 8.10 The triangular model of a 1 × 8 latching switch.

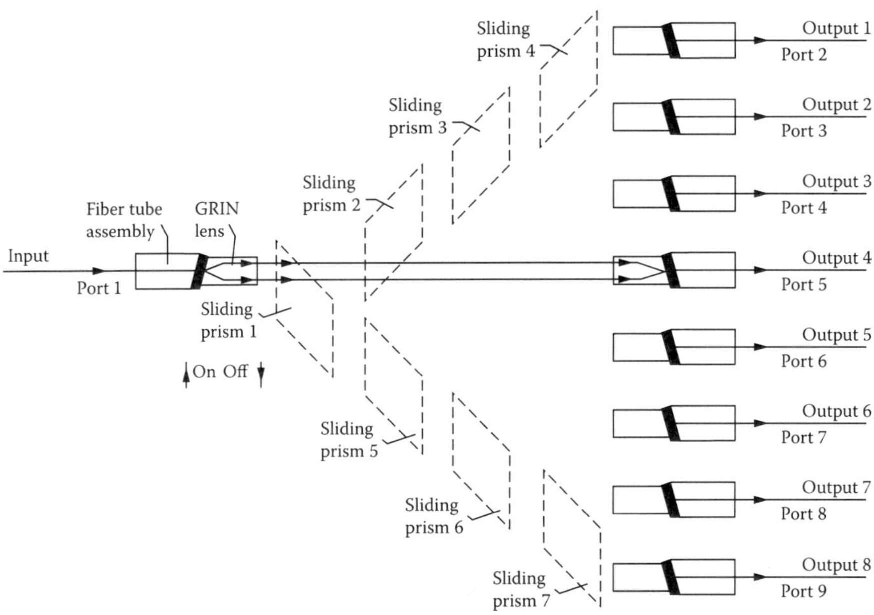

FIGURE 8.11 Signal from port 1 to port 5 in a 1 × 8 latching switch.

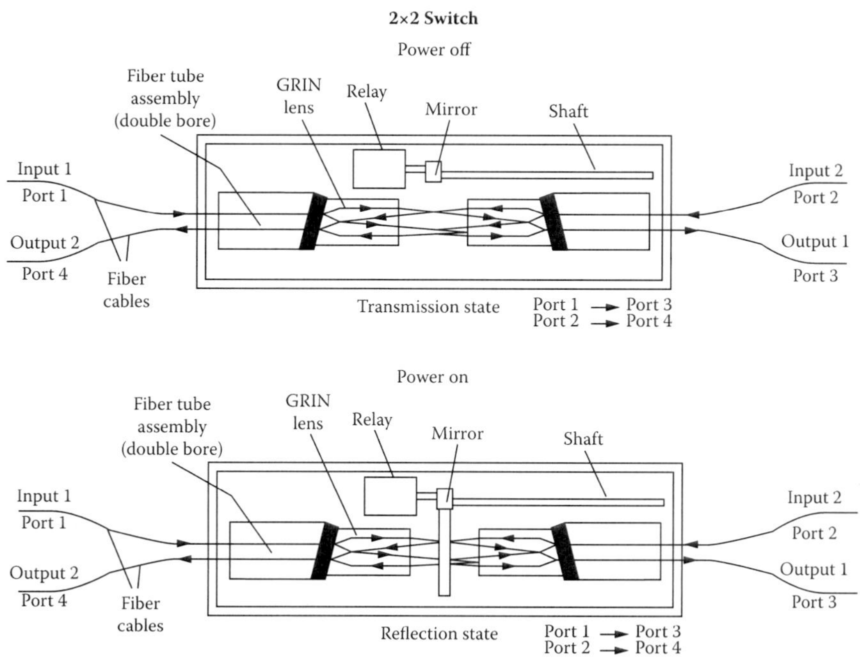

FIGURE 8.12 A 2 × 2 optomechanical switch.

8.3 ELECTRO-OPTIC SWITCHES

Switches with no moving parts can be built by using some of the passive devices, such as Mach-Zehnder interferometers (MZIs) and couplers. Some optical materials, such as lithium niobate crystal ($LiNiO_3$), Avalanche Photo Diode (APD) ($NH_4H_2PO_4$), and KDP (KH_2PO_4), exhibit an electro-optic effect. The index of refraction (RI) of the optical material changes in the presence of an electric field. These optical materials are used in building devices, such as MZI, APD, and KDP. An electric field applied across the lithium niobate crystal causes a variation in the index of refraction. This changes the transit time, creating a phase shift of the optical signal passing through the lithium niobate crystal.

Mach-Zehnder interferometers (MZIs) are used in building optical devices, which are used in a wide variety of applications in optical communication systems. The basic requirement of the Mach-Zehnder interferometer is to have a balanced configuration of a splitter and a combiner connected by a pair of optically matched waveguides, as shown in Figure 8.13. The optical signal entering the Mach-Zehnder interferometer input port is split through a Y-splitter section into two equal components. Each component goes to one of the two arms of the Y-splitter. When there is no phase change in signal components after passing through both arms of the interferometer the signal components are recombined at the Y-coupler immediately before the optical signal exits the Mach-Zehnder interferometer. The recombination of the two signal components takes place as constructive interference between

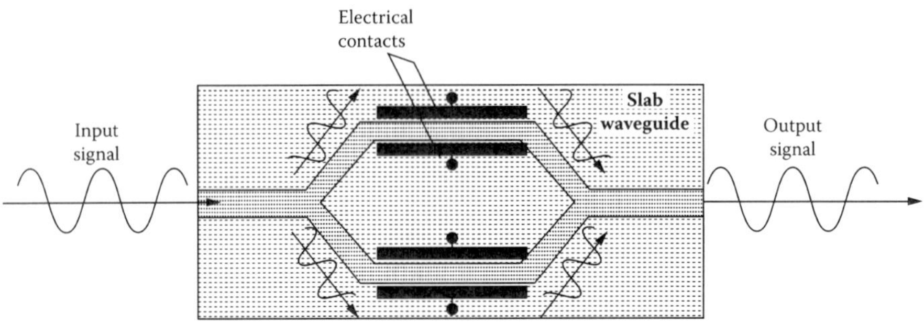

FIGURE 8.13 Mach-Zehnder interferometer acts as a passive device.

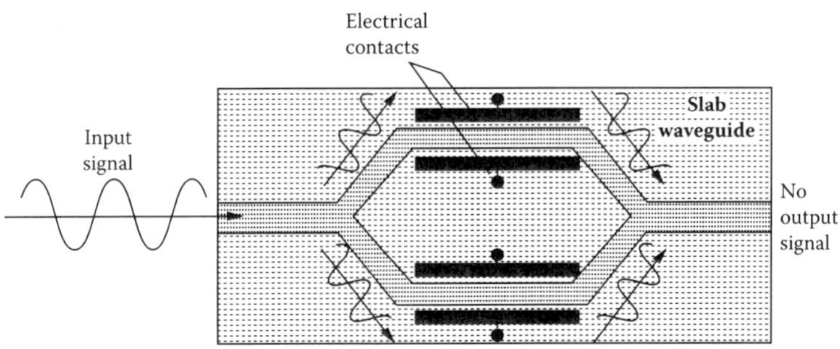

FIGURE 8.14 An electro-optic switch using a Mach-Zehnder interferometer.

two components and regenerates the original optical signal. In this case, the Mach-Zehnder interferometer acts as a passive device.

When an electric field is applied to one arm of the Mach-Zehnder interferometer, the index of refraction changes and causes a 180° shift in the phase of the signal component due to the change in optical path length of this arm. As shown in Figure 8.14, when there is a difference in phase at the destination Y-coupler, the signal components will be out of phase with one another. The signal components' recombination will be lost because the components will cancel each other out in destructive interference. If the phase difference is a full 180°, then the output will be zero. In other words, applying the electric field to one of the arms of the Mach-Zehnder interferometer will then make the phase shift one of the signal components. In this case, the Mach-Zehnder interferometer acts as an active device, since an applied electric field causes the switching.

Using the same principle as discussed above, you can build an electro-optic switch using two branching waveguides arranged like a 3 dB coupler to switch one input between two outputs. You can replace the one input with two parallel outputs coupled to the pair of switching waveguides by a combining coupler, as shown in Figure 8.15. An electric field applied to one arm of the waveguide causes a 180° shift in the phase of the signal component. The electrical voltage is raised or lowered to shift the delay between waveguides by 180°. This directs the output from one waveguide on the right side to the other output. Because signal interference depends on

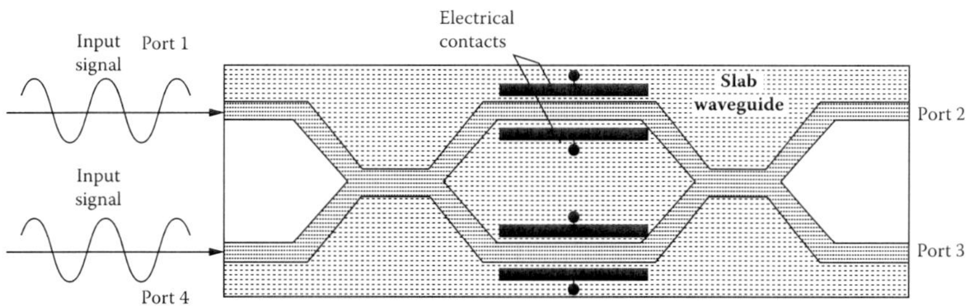

FIGURE 8.15 A 2 × 2 electro-optic switch.

TABLE 8.1
Input and Output Signals Connections

Voltage	Connections
V_1	Port 1 to port 2
	Port 4 to port 3
V_2	Port 1 to port 3
	Port 4 to port 2

phase shift, it is possible to further increase the voltage to switch the signal back to the other output. Table 8.1 presents the possible outcomes achieved by applying different voltages across the waveguide arms.

8.4 HIGH-SPEED OPTICAL SWITCH USING A FIBER OPTIC COUPLER

A high-speed all-optical switch using a fiber optic coupler and a light-sensitive variable index material is illustrated in Figure 8.16. Refractive index variation with light is the principle of the switch operation. Evanescent wave coupling between two monomode fibers is extremely sensitive to the index of refraction of the material surrounding the coupling region. Two ground and polished fibers, producing an evanescent field, can be brought in close proximity so that light in one fiber will couple into the other fiber in any desired ratio. Such polished couplers have been constructed to produce very low losses.

A similar coupler may be made by timed etching of the fibers. Hydrofluoric acid may be used to remove as much cladding as desired; this exposes the core and produces an evanescent field. Within the etched regions, fiber-to-fiber optical coupling will occur for fibers placed in close proximity. The coupling efficiency will depend on the index of refraction of the surrounding medium, the core-to-core separation, the length of the interaction, and the amount of etching.

Another evanescent field coupler is based on a nonetched, fused, and drawn coupler. This type of coupler is drawn to such an extent that the core is essentially

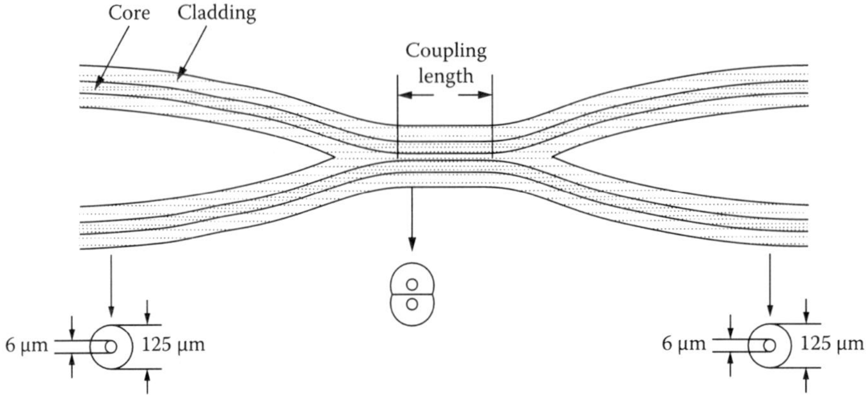

FIGURE 8.16 A high-speed optical switch using a fiber optic coupler.

lost and the cladding reaches a diameter near that of the undrawn core. The claddings become the core, and the evanescent field is forced outside the new core (waist region) into the air. Twisting two fibers together and fusing by using heat makes fused biconical tapered couplers.

The contribution of charge carriers (electrons) to the index of refraction provides a simple way to modulate the index by the introduction or withdrawal of such carriers in the material. This can easily be done by the creation of electron-hole pairs if the material is also photoconductive (optical modulation).

One of the important parameters of the couplers is the coupling length. The coupling length is wavelength dependent. Thus, the shifting of power between the two parallel waveguides will take place at different places along the coupler for different wavelengths. Figure 8.17 shows two wavelengths entering at ports 1 and 2. When coupler length is made exactly to match the wavelength of the signal, the coupler works to combine wavelengths. Combined signals exit from port 2.

Figure 8.18 shows the reverse process where two different wavelengths arrive on the same input fiber at port 1. At a particular location along the coupler, the wavelengths will be in different waveguides. Then the wavelengths separate exactly, and each wavelength exits from a different port. In this case, one wavelength exits

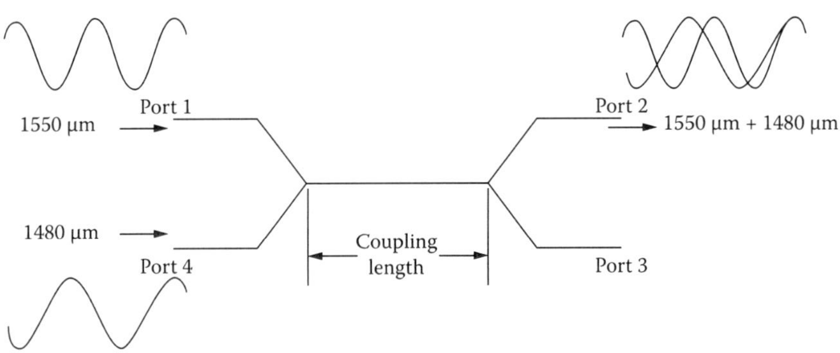

FIGURE 8.17 Two wavelengths entering at ports 1 and 2.

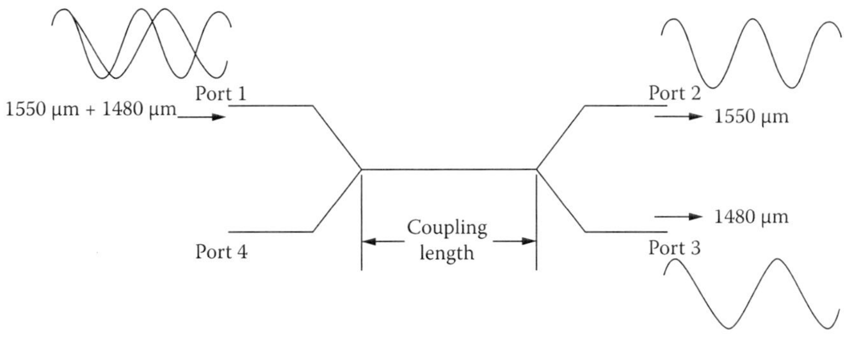

FIGURE 8.18 Two wavelengths entering at port 1.

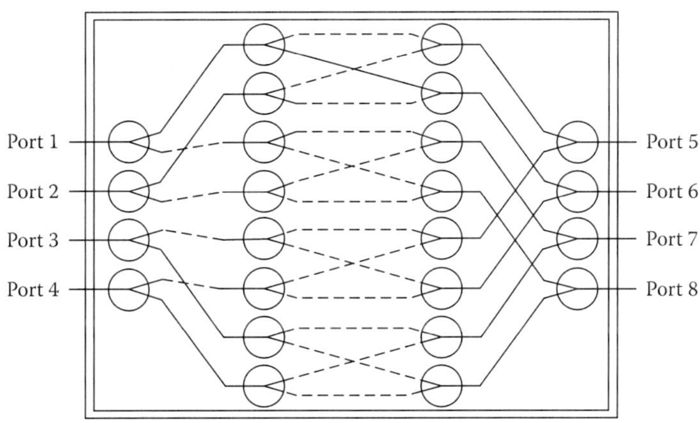

FIGURE 8.19 A 4 × 4 optical space-division switch.

from port 2 and the other from port 3. The processes described in Figures 8.19 and 8.20 are performed in the same coupler. This process is bidirectional. The coupler in Figure 8.19 works as a splitter; the same coupler in Figure 8.20 also works as a combiner.

There are other types of switches that use the same elements of either micro-optomechanical switches or microelectro-optical switches. They employ couplers for switching between the inputs and outputs. These types of optical switches are used in communication systems. Figure 8.19 shows the configuration of a 4 × 4 optical space-division switch. The switch is designed to connect any input port to any output port as desired by the user. Any input may be switched to any output. However, two inputs may not go to the same output at the same time. The device is bidirectional such that once a connection has been established between an input port and an output port, that particular connection may be used in either one or both directions. The switches have no moving parts, are very stable, and are reliable while exhibiting very low loss, thereby reducing the need for expensive amplifiers.

Figure 8.20 shows the cross-connect switch, which selects outputs by optical cross-connecting. This results in a significant reduction of overall complexity and

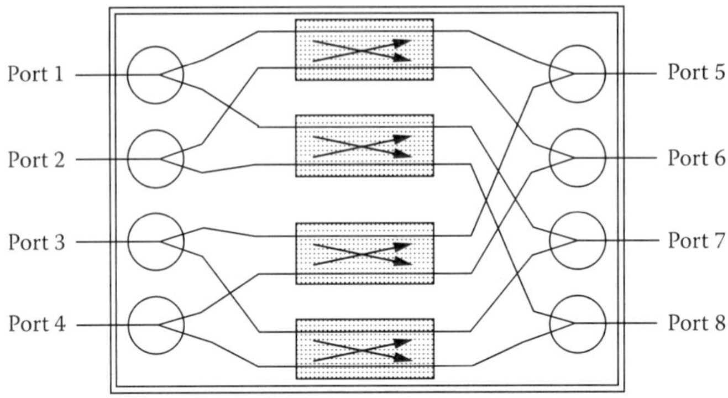

FIGURE 8.20 A cross-connect switch.

number of required elements. These switches are used in protection/backup switching, optical cross-connecting, network testing and monitoring, optical routing, and optical burst switching. A 4 × 4 switch configuration can be cascaded to build 16 × 16 to 256 × 256 switch configurations.

8.5 ACOUSTO-OPTIC SWITCHES

Sound waves are generated when a material is in a mechanical vibration mode. They can also be generated by acoustic transducers. Like any light wave, a sound wave is a moving wave that has a frequency. Light waves travel at the speed of light; sound waves travel at the speed of sound, which is much slower than the speed of light waves. Sound waves are used to control light transmission in acousto-optic switches and modulators. The refractive index of some optical material is altered by the presence of sound waves. The sound wave causes regular zones of compression and tension within the optical material. This creates a regular pattern of changes in index of refraction n of the optical material; this is called a Bragg diffraction grating. Within the optical material, there is interference between sound and light waves. The power of the deflected light is controlled by the intensity of the sound wave. The angle of deflection is controlled by the frequency of the sound wave. Figure 8.21 illustrates a design of an acoustic-optic switch. The figure shows that an incident light can be controlled by the frequency of the sound wave. The incident light exits the switch from one or more selected output ports, depending on the sound wave intensity.

8.6 MICROELECTROMECHANICAL SYSTEMS (MEMS)

Microelectromechanical systems (MEMS) are a rapidly growing technology for the fabrication of miniature devices using processes similar to those used in the integrated circuit industry. MEMS are widely used in optical switching in telecom networks. The appeal of MEMS goes beyond just switching applications in defense, aerospace, and medical industries. MEMS technology provides a way to integrate electrical, electronic, mechanical, optical, material, chemical, and fluids engineering on very small devices that range from a few microns to 1 mm. MEMS devices have

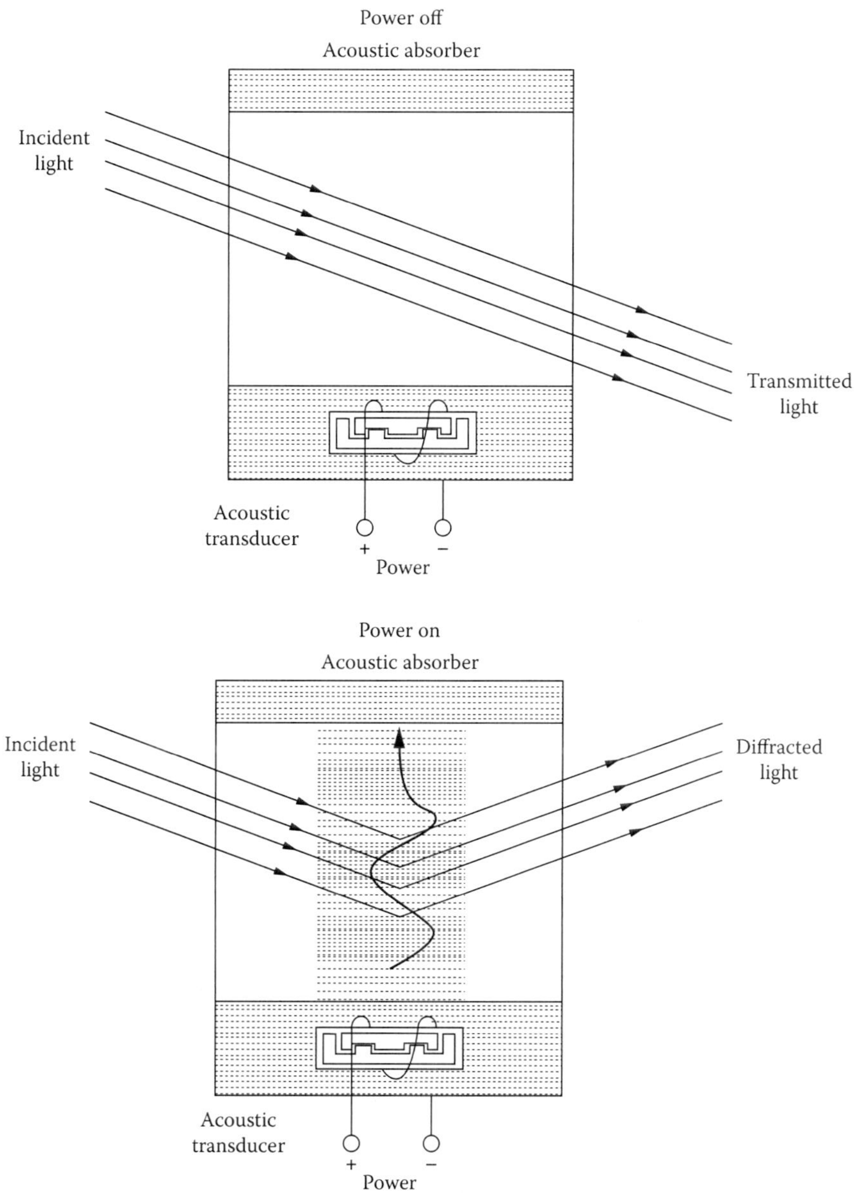

FIGURE 8.21 A design of an acoustic-optic switch.

many important advantages over conventional optomechanical switches. First, like integrated circuits, they can be fabricated in large numbers, so that cost of production can be reduced substantially. Second, they can be directly incorporated into integrated circuits, so that far more complicated systems can be made than with other technologies. Third, MEMS have a small size, a low cost, and high reliability and stability. Fourth, they have the important capability of high-density digital transmission communication with different bandwidths.

There are two categories of MEMS switches: MEMS 2D and 3D. They are typically fabricated onto a substrate that may also contain electronics needed to drive the MEMS switching element. MEMS-based optical switches route light from one fiber to another to enable equipment to switch traffic completely in the optical domain without requiring any optical-to-electrical conversion. At the core of MEMS 2D matrix switches is an array of micro-mirrors capable of redirecting light either in free space or within a waveguide framework. The 2D switch architecture shown in Figure 8.22 employs one mirror for every possible switched node in a matrix switch, and thus requires N^2 mirrors for an $N \times N$ array. The 2D mirror arrays are characterized by two-state mirror positioning. One state is inactive and requires only that the mirror can be parked out of the optical path. During the switching state, the mirror redirects the light path. Mirror positioning accuracy, repeatability, and stability are critical in determining switch performance. Unlike 3D switch architectures that require servo positioning of individual mirrors, a 2D switch can rely on passive positioning control of the switched mirror, simplifying the control scheme. But a successful MEMS 2D approach must provide means for actuating the mirror into a highly predictable, stable state and hold it there indefinitely.

MEMS use an array of pop-up MEMS mirrors fabricated on the surface of a silicon wafer. The mirror is hinged to allow its rotation off the plane of the substrate to an angle of 90°, where it redirects a light channel from the through to the cross-state, as shown in Figure 8.23. An addressing scheme is required to select individual mirrors for actuation into the popped-up state and also for positioning them with sufficient accuracy for efficient coupling into the switched channel. The 2D MEMS array, described here, called Mag O × C (which stands for magnetically optical actuated cross-connect), uses a combination of magnetic and electrostatic actuation to rotate the mirrors and select and deselect individual mirrors for clamping into the UP or DOWN state.

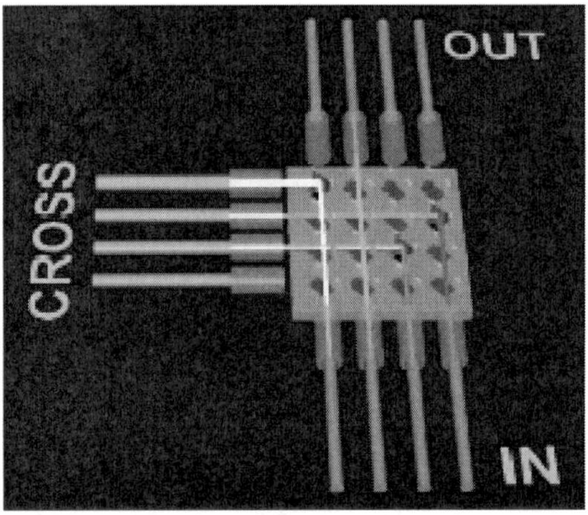

FIGURE 8.22 MEMS 2D switch architecture.

FIGURE 8.23 Pop-up mirror array with passive mirror alignment.

To rotate unclamped mirrors into the UP state, magnetic actuation is implemented globally by applying an external field generated with a small electromagnet. The magnetic signal only needs to be applied momentarily. Using a global field avoids the need to fabricate individual magnetic actuators for each chip. Mirrors are fabricated with a layer of attached nickel to produce torque on the mirror hinge in response to the applied field. The nickel plate aligns with the magnetic field lines and generates a magnetostatic torque on the mirror. This lifts the mirror off the substrate and orients it near the desired vertical position, where electrostatic force can take over in setting and holding the final desired mirror position. An electrostatic field is applied mirror by mirror, either to hold the mirror down against the torque produced by the magnet or to hold the mirror in the UP position against the restoring force of the elastic hinge. Since the magnetic field is applied globally, all mirrors will attempt to rotate when the field is turned on. Only the mirrors to be rotated into the UP position are unclamped; all others are held down electrostatically. Similarly, mirrors clamped in the UP state remain so until the electrostatic signal is removed; the magnet is no longer needed to hold the mirror up. The combination of magnetic and electrostatic actuation provides an effective means for configuring a mirror array without resorting to complex individual actuators for each mirror. Since electrostatic clamping of the mirrors requires virtually no current flow, the switch array consumes very little steady-state power. Power is consumed only during transitions when the magnet is activated. The components of the switch are packed in a packaging base and lid.

8.7 3D MEMS-BASED OPTICAL SWITCHES

A 3D MEMS-based optical switch routes light from any of 80 input fibers to any of 80 output fibers. Designed for fiber-based test and measurement, 10 Gbit/s Ethernet, high-definition video, and telecom applications, this all-optical microphotonic subsystem fits in the palm of a hand. The switch design is based on 3D microelectromechanical system (MEMS) mirror arrays, which is called reflection. It can switch

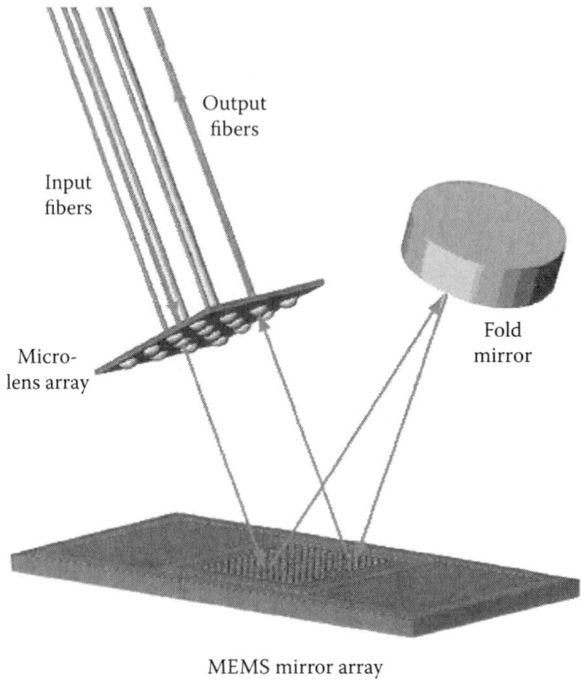

FIGURE 8.24 A 3D MEMS optical switch architecture.

signals within 10 ms, which is well within the telecommunications requirements for communications applications. 3D designs have switching elements accommodating hundreds, even thousands, of ports. Figure 8.24 shows a 3D MEMS optical switch. The design of the 3D MEMS is simple, solves mechanical and optical issues, is easy to fabricate, and achieves manufacturing tolerances that are accepted by the telecommunications industry.

A 3D MEMS design is shown in Figure 8.25. A micro-mirror array rests at the top a single piece of silicon on a ceramic substrate. No bonding pads or other integrated electronics exist on the chip. All routing to the rest of the actuation electronics is done on the back end of the ceramic substrate. Additional electronics are located on

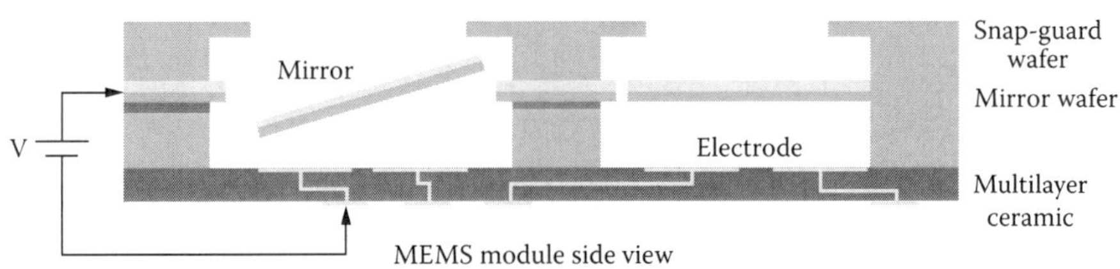

FIGURE 8.25 A 3D MEMS optical switch design.

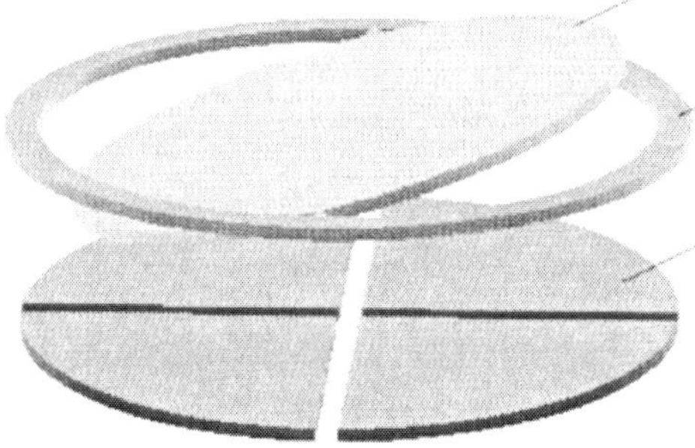

FIGURE 8.26 A torsional micro-mirror.

a photodetector card, constant-delay two-pole Bessel filters, and analog-to-digital converters with a conversion-phasing time. Parallel-plate electrostatic actuations of the mirrors, with potentials of about 200 V, are provided by high-voltage linear amplifiers.

Figure 8.26 shows a torsional micro-mirror, which is driven by a vertical bidirectional comb driver (microelectrostatic actuator). Underneath the mirror plate is the substrate electrode. During operation, a voltage is applied to the electrode in order to generate an electrostatic attractive force on the mirror plate. The mirror plate rotates around the supporting axis in two dimensions. Such tilting of the mirror position can direct light from many distinct input ports to any of many distinct output ports.

8.8 MICRO-OPTOMECHANICAL SYSTEMS (MOMS)

On-chip integrated micro-optomechanical systems (MOMS) were developed for a variety of applications for optical telecommunications. It is a new technology that allows for the integration of multiple passive and active components at the chip level. The technology is an extension of integrated optics and is based on suspended waveguides fabricated on chips, which are integrated with other optical components. It is used for a variety of optical solutions, including optical switching and cross-connect, signal dispersion correction, configurable optical add/drop multiplexers, and signal intensity equalizers. The technology bases its main technological concept on integrating optics at the chip level. The technology explores low-cost, high-performance, and planar optical waveguide switches. Waveguides are used to channel light, rather than allowing it to propagate in free space. The use of waveguides allows a degree of freedom in reducing size while at the same time operating in a controlled physical environment. Switching time and losses are very low. By using waveguides, photons can be channeled in a controlled fashion, making it possible to lay out a photonic network within the chip with low losses. Since silicon is used as the propagating material, wavelength transparency for 1,300 nm to 1,600 nm is utilized. The use of silicon for the waveguide material creates a tight confinement of light that allows the use of very small curvature radii, enabling a significant reduction of footprint chip size.

9 Continuous Improvement in a Production Line

9.1 INTRODUCTION

Most companies and businesses around the world operate in a very competitive marketplace, telecommunication and car companies. Companies such as Bell Canada, Rogers, etc., are competing directly against each other to offer customers cheap rates and good services. In order to ensure they stay healthy in the market, the companies continually improve their products and customer service. This process is called continuous improvement. The continuous improvement process is to improve their manufacturing system and product and to create opportunities for further improvement. This is usually achieved by improving the efficiency of the manufacturing processes, reducing waste, training employees, and investing in new up-to-date technology.

9.2 UNDERSTAND THE NEED FOR CHANGES

There are many steps that can be adopted to improve the product and reduce costs, which include:

1. Improving workplace environment.
2. Improving safety of the employees.

3. Improving production line operations.
4. Reducing handling time.
5. Reducing process time.
6. Using high-quality alternative materials.
7. Increasing quality of the products.
8. Eliminating rejects.
9. Applying 5S's.
10. Reducing costs in any modifications and changes.
11. Improving employee moral.
12. Improving delivery performance.
13. Reducing excessive inventories.
14. Eliminating equipment breakdown.
15. Updating calibration and software.
16. Using online scheduling for orders, employees, and inventory.
17. Eliminating process bottleneck.
18. Improving capacity of the production line.
19. Creating R&D.
20. Applying lean manufacturing wherever it is possible.
21. Applying automation process wherever it is possible.
22. Creating the foundations for future planning:
    a. Employee training.
    b. Adapting new technology, instruments, and tools.
    c. Investing in research and development activities.
    d. Increasing the revenue of the company.
    e. Increasing customer confidence.
    f. Increasing job security.
    g. Increasing productivity.
    h. Reducing manufacturing cost.

Figure 9.1 shows a summary of the support that is given to a production line from all engineers, supervisors, managers, R&D, support staff, and marketing.

9.3 PREPARING FOR THE CHANGE

Meeting the global competition requires change. An organization must be prepared for such necessary change. Quality improvement must be instituted in every aspect of everything that organization does in daily and future activities. If the employees are better prepared for change, then positive changes can be achieved. The requirements for getting an organization ready for the drastic change that may be needed for quality improvement include:

1. Keep everyone informed of the impending changes.
2. Get all employees involved.
3. Make employees a part of the decision-input mechanism.
4. Highlight the benefits of improved quality.
5. Train employees on their new job requirements.

FIGURE 9.1 Summary of the support to a production line.

6. Apply a new advanced technology in manufacturing, scheduling, orders, delivery, and planning.
7. Create an environment for job enrichment.
8. Allay fears about loss of jobs because of improved quality.

9.4 HANDLING TIME

Handling time is the time in handling material from one person to another in a production line. Handling time is the time wasted during the manufacturing process due to unnecessary movement of material or product or due to inefficient use of the operator's time. It reduces production and increases cost.

The following advantages of reducing handling time are:

1. It reduces cost.
2. It increases productivity.
3. It saves time.
4. It improves delivery performance.

5. It gives the benefit of better visibility of problems, reducing frustrations through better organization.
6. It increases the revenue of the company.
7. It increases customer confidence.

The following improvements reduce the handling time:

1. Floor space utilization
2. Excessive storage space on the production line floor and workstations
3. Utilizing employees' time efficiently
4. Excessive nonconforming materials in the production line workstations
5. Manufacturing processes improvement
6. Employee training
7. Adapting new handling technologies, instruments, devices, systems, and tools

9.5 REDUCING PACKAGING AND SHIPPING TIMES

Reducing packaging and shipping times is very important in the manufacturing of any products. In the photonics industry, proper packaging is required all year long because the products shipped are very expensive to replace in case of damage during transportation. Packaging and shipping processes are at the end of the production line. The products must be packed in a proper process. Packaging consumes process and handling times. Packing processes include the title, labels, bar code, log, warning, and customer address being fixed on a shipping box. Each product has its own packing and shipping processes.

9.5.1 Packaging Process

Starting with the outside of the package, usually a cardboard box of some kind and size can be chosen to accommodate the product. Normally, a good 5 cm on each side of extra clearance room is allowed in the packaging box. This extra room is for the cushioning materials used to protect the product inside the box. Each packing box has certain dimensions. The packaging boxes are delivered in many ways, depending on the size of the boxes. Delivery is normally carried out by trucks. Every square centimeter of a truck space is valuable and costs money. Therefore, the size of the packaging box depends on the size, type, and sensitivity of the device.

Some packaging materials are used inside the box to protect the product during transportation. Cushioning materials are used to fill up the space around the product. These cushioning materials can be hard and soft. The cushioning materials type, size, and specifications depend on the product sensitivity and shipping destination. The cushioning materials can be bubble wraps, packing peanuts, blank newspapers, kraft papers, Styrofoam, plastics, cardboard sheets, electrostatic bags, polyethylene form sheets, etc. When fragile items such as glass, prisms, mirrors, and other optical items can be shipped using a good method to package them, it is by a combination of bubble wrap and peanuts, as shown in Figure 9.2(a). A polyethylene form sheet can also be added in between the items, as shown in Figure 9.2(b).

(a) Plastic sheets (b) Polyethylene form sheet

FIGURE 9.2 Cushioning materials are used between items.

Packing peanuts can be used in the packaging of a device. The normal procedure is to create a bed for the device on the bottom of the packaging box. Then the device can be placed in the center of the box. The rest of the void space in the box can be filled with more peanuts, making sure that the peanuts settle into place around and cover the device. In this case, the device is secured inside the box and surrounded by the peanuts and protected from any damage during shipping. Sometimes, crumpled newspapers or kraft paper can also be used to fill any space inside the box.

For static sensitive devices, a common strategy is used to protect the devices from damage by electrostatic charge. Such a device is usually wrapped with materials that do not allow charges to go through it. For example, electrostatic discharge bags come in silver or pink color, in different sizes, as shown in Figure 9.3. Silver electrostatic discharge bags are for maximum protection, and pink bags are for less protection. For very sensitive devices, all the packed boxes can be wrapped with antistatic film to prevent any electrical charge leaks.

(a) Silver bags (b) Pink bags

FIGURE 9.3 Electrostatic discharge bags.

Some packaging materials have good benefits, such as:

1. Excellent protection and prevention against scratch
2. Moisture resistance
3. Light weight
4. Good for wrapping individual items
5. Easy to use
6. Recyclable

The following improvements reduce the packaging and shipping times are:

1. Employee training
2. Adapting new technology, instruments, devices, and tools
3. Packaging and shipping processes improvement

9.5.2 Shipping Labels and Tags

There are many types and colors of international shipping labels and tags that are used in the shipping of different devices. Figure 9.4(a) shows labels that are used on shipping boxes to indicate that the materials are fragile and should be handled with care. Such materials are glassware, sensitive electronic devices, lenses, photonic measurement devices, etc. Figure 9.4(b) shows special safety labels and information labels, and Figure 9.4(c) shows a label indicating the box can be recycled after use.

9.5.3 Sealing the Shipping Box

Before sealing the shipping box, a copy of the contents, including shipping information, inventory, instruction manual, and other documents, should be supplied. The address of the customer must be clearly labeled on the box according to the shipping procedure of the company or as per customer requirements. Use plastic tape to seal the box. The tape should be environmentally friendly as well as durable to hold the box together.

A copy of the invoice is attached to the outside of the box. The invoice has the shipping and receiving information, a list of items, a destination address, a shipping date, the shipping company, etc.

9.6 5S IMPLEMENTATION

The steps of the 5S's is:

1. Sort
2. Set in order
3. Shine
4. Standardize
5. Sustain

5S is the key first step in workplace improvement through an emphasis on a work environment that ensures “a place for everything, and everything in its place, clean,

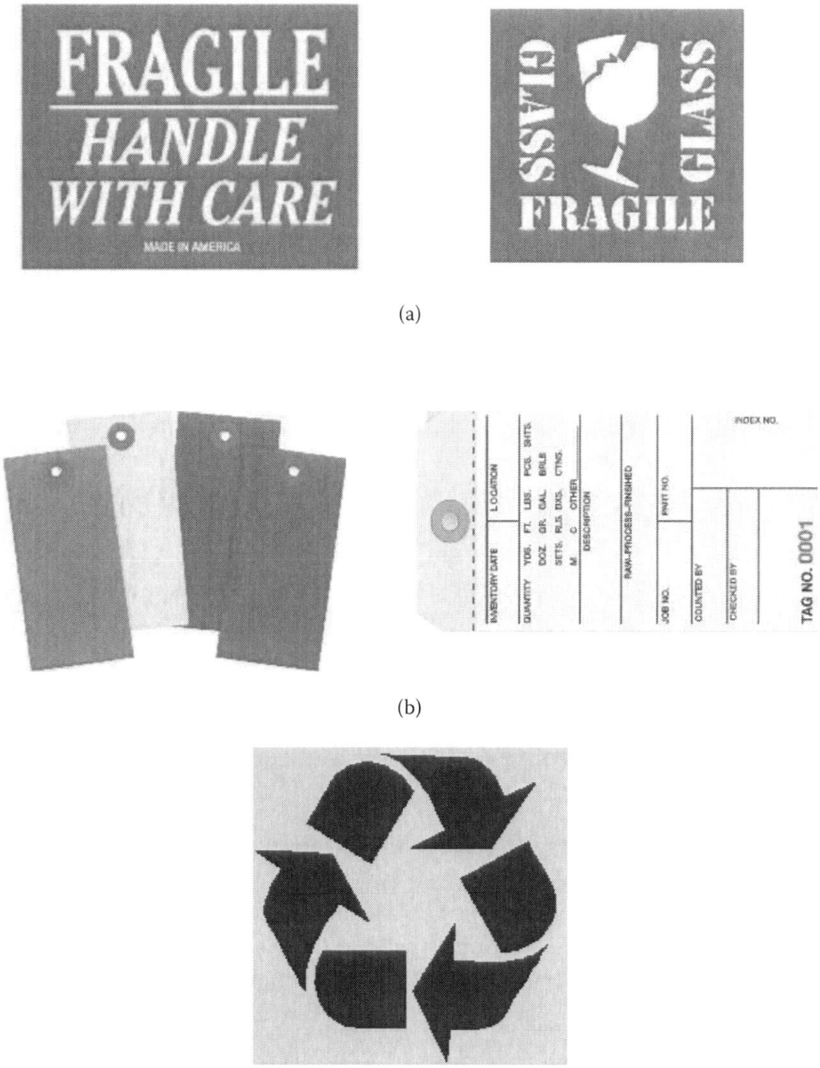

FIGURE 9.4 Some shipping labels and tags.

organized, and ready to use." 5S is not just a cleanup campaign. It is a system that allows employees to work more efficiently.

5S requires perseverance, determination, and ability to pay attention to small details of the production line (parts, stations, materials, equipment, instruments, and employees). 5S gives the benefit of quick visibility of problems, reduces frustration through better organization, and improves efficiency, effectiveness, and safety. A detailed description of the 5S's is in Chapter 11.

9.7 MODIFICATIONS OF A PRODUCTION LINE AND PRODUCT

It is necessary to consider the following steps in the modifications of a production line and product:

1. Modify the manufacturing processes.
2. Check and inspect incoming raw materials.
3. Quality control processes on each manufacturing step (sometimes check, test, or inspection).
4. Use alternative raw materials in building a device to reduce the raw materials costs and manufacturing steps.
5. Modify the floor plan.
6. Improve employee training and education (training the employees to create a good working environment, reduce rejects, and increase productivity).
7. Check the working environment of the employees (add proper lighting systems, safety shoes, goggles, prescription glasses, healthcare, holidays, celebrations, parties, salary increases, awards, etc.).
8. Update jigs, tools, devices, systems, ovens, computers, etc.
9. Regularly calibrate measuring and test equipment and devices.
10. Propose new jigs.
11. Check air pressure, air quality, and air-conditioning systems.
12. Update all manufacturing processes and adapt a new technology whenever it is possible.
13. Calculate the process, test, handling, waste, packaging, etc., times with the employees' contribution.
14. Update the delivery process of a complete device to customers.
15. Apply 5S's on the production line as routine practice.
16. Create short and long maintenance plans.
17. Apply lean manufacturing processes where possible and applicable.
18. Adopt any other steps that improve the products and production line to reduce time, increase productivity, reduce rejects, reduce cost, and achieve customer satisfaction.

9.8 REDUCING WASTE

Waste types and reducing waste in a production line are presented in Chapter 10. The following are the types of waste in many production lines. Elimination of these types will reduce the cost of products.

1. Waste of inventory
2. Waste of material movement
3. Waste of overproduction
4. Waste of inefficient process
5. Waste of defects

6. Waste of waiting
7. Waste of motion
8. Waste of person

9.9 SOLVING PROBLEMS IN MANUFACTURING LINES

There are many problems that occur in production lines. These problems vary from one manufacturing line to another. The problems' severity depends on the product type and quantity. Production lines experience the following problems:

1. Long manufacturing cycle times
2. High ratios of motion to work
3. More cost-added than value-added activities
4. Worker and machine specialization
5. Poor visibility of problems
6. Need to appropriate jigs
7. Delayed problem resolution
8. Lack of employee concentration and skills
9. Inefficient use of floor space
10. Ineffective regular and scheduled maintenance plans
11. Inaccurate manufacturing processes
12. Large quantities of rejects
13. Huge quantities of work-in-progress
14. Poor follow-up of the new technologies
15. Unfortunate management
16. Inefficient R&D activities

9.10 JIG USAGE IN PRODUCTION LINES

In workshops and manufacturing lines, a jig is a type of custom-made tool used to hold a work piece and control the location or motion of another tool. The primary purpose of using a jig is to provide repeatability, accuracy, and interchangeability in the manufacturing of a specific product. Jigs are made to increase productivity through consistency, to do repetitive activities, or to do a job more precisely with reduced manufacturing time. There is also the use of a fixture in the production lines. The fixture is used when the work piece is in a fixed position. There is no difference between the two definitions. Some types of jigs are also called templates or guides. A jig is a tool/device that does a specific job in the production lines. For example, jigs include machining jigs, woodworking jigs, welder jigs, jeweler jigs, lighting jigs, etc. Some jigs can be used in other production lines to do multiple jobs. There are many examples, such as positioners, holders, posts, etc.

Figure 9.5 shows a new jig is built by the production support section. The jig is used to supply light into fiber optic connector tests. The cost of the jig is significantly lower than the cost to build and operate. The jig works in these tests efficiently.

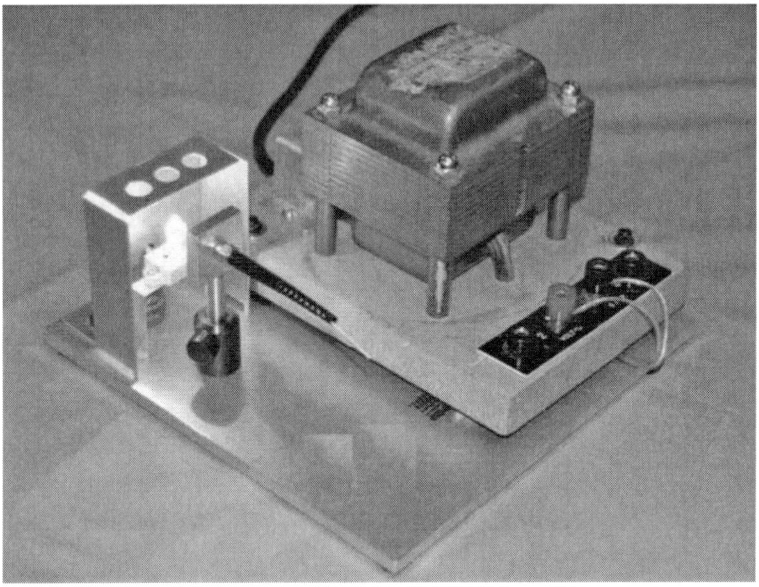

FIGURE 9.5 A jig to supply light into fiber optic connector tests.

9.11 CONTINUOUS IMPROVEMENT SCHEDULE

The steps in continuous improvement are presented in Figure 9.6, are as follows:

1. Document reality.
2. Identify waste.
3. Plan countermeasures.
4. Reality check—does this work in the real world?
5. Make changes.
6. Verify change—evaluate.
7. Measure results.
8. Make the improvement the standard.
9. Celebrate success.
10. Repeat.

9.12 STANDARD WORK IMPROVEMENT PROCESS

The following are steps in a standard work improvement process:

1. Standardize the job and adhere to it.
2. Expose problems and waste.
3. Solve problems and eliminate waste.
4. Document solutions and update processes.
5. Employee training.
6. Implement solutions.

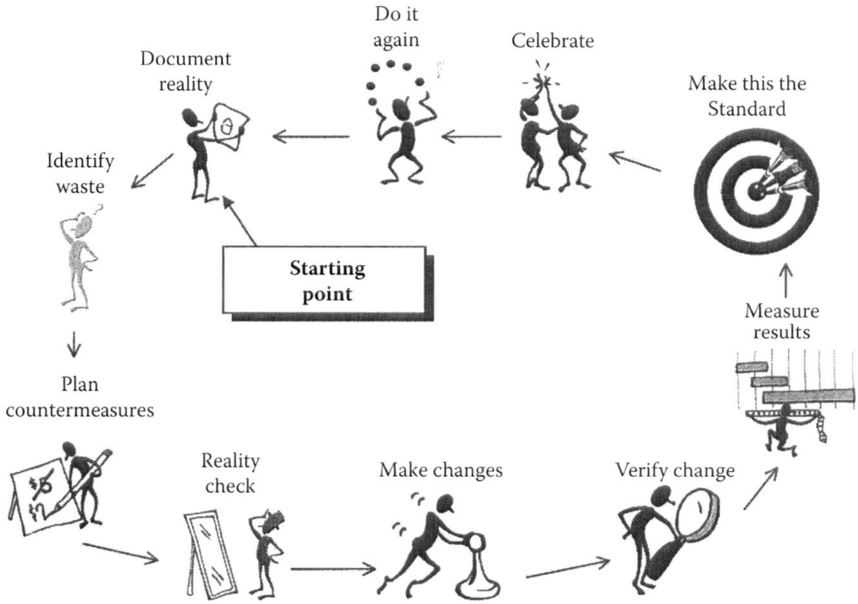

FIGURE 9.6 Continues improvement cycle.

9.13 CONTINUOUS FLOW PRODUCTION (CFP)

The current marketplace for manufacturers is dynamic, global, and consumer-focused. Shorter product life cycles, reduced profit margins, and consumer demands for more variety in products are forcing changes in the way products are manufactured. One of the approaches that modern manufacturers implement is continuous flow manufacturing. Continuous flow means that items are processed and moved directly to the next process one piece at a time. Each processing step completes its work just before the next process needs the item.

9.14 BASIC ELEMENTS OF CONTINUOUS FLOW PRODUCTION

The process of continuous flow production is an integration of the continuous flow distribution theme with reduced inventories and throughput times within the manufacturing process to support the continuous flow distribution process. Continuous flow manufacturing encompasses four basic elements:

1. Based upon customer requirements, an overall manufacturing network must be configured.
2. Manufacturing requirements are identified, and strategic master plans are developed and implemented for each operation.
3. Information and management systems for the manufacturing process and operations are assessed, defined, purchased, and implemented. The information systems will drive the manufacturing continuous flow process and

will interface with all of the organization's business systems. Effective information systems would contribute to reductions in errors, maximization of labor utilization, maximization of equipment utilization, improved customer service, and ultimately market domination.

4. Once the manufacturing network, manufacturing requirements, and information and management systems are in place, the process of continuous improvement must be installed.

9.15 JUST-IN-TIME (JIT)

Just in time (JIT) is an inventory strategy implemented to improve the return on investment of a business by reducing in-process inventory and its associated costs. The process is driven by a series of signals, or kanban, that tell production processes to make the next part. Kanbans are usually simple visual signals, such as the presence or absence of a part on a shelf. When implemented correctly, JIT can lead to dramatic improvements in a manufacturing organization's return on investment, quality, and efficiency.

JIT is frequently known as a pull system, as everything is pulled in response to actual demand. Demand forecasts are not used in kanban systems. This is the opposite of the traditional push manufacturing philosophy, in which everything is made to forecasted future needs. Push systems often encounter serious difficulties when demand forecasts turn out to be inaccurate.

A JIT system is an essential part of continuous flow production, which has a lot of benefits contrary to the push system. Here are six main benefits of a pull system:

1. Setup times are significantly reduced in the warehouse.
2. The flow of goods from warehouse to shelves is improved. Having employees focused on specific areas of the system will allow them to process goods faster.
3. Employees who possess multiple skills are utilized more efficiently. Having employees trained to work on different parts of the inventory cycle system will allow companies to use workers in situations where they are needed when there is a shortage of workers and a high demand for a particular product.
4. Better consistency of scheduling and employee work hours. If there is no demand for a product at the time, workers don't have to be working.
5. Increased emphasis on supplier relationships. Having a trusting supplier relationship means that a company can rely on goods being there when it needs them in order to satisfy the company and keep the company name in good standing with the public.
6. There are many hidden manufacturing problems associated with the buildup of inventory. Reducing inventory to the minimum in the JIT system brings those problems to the surface without delays.

9.16 CHARACTERISTICS OF THE FLOW PRODUCTION

Continuous flow production is aimed to achieve the shortest possible cycle time by eliminating waste, using dependent demand scheduling, as opposed to independent demand. It is derived from the Toyota Production System, and its key thrust is to increase the value-added work by eliminating waste and reducing incidental work. The characteristics of flow production are:

1. Straight and short product flow patterns
2. Make to order
3. Single-piece production
4. Just-in-time materials/pull/dependent demand scheduling
5. Short cycle times
6. Highly flexible and responsive processes
7. Highly flexible machines and equipment
8. Quick changeover
9. Continuous flow work cells
10. Collocated machines, equipment, tools, and people
11. Compressed space
12. Multiskilled employees
13. Empowered employees
14. High first-pass yields with major reductions in defects

Flow production is in direct opposition with traditional mass production approaches characterized by use of economic order quantities, high-capacity utilization, and high inventory. The demand-based pull of material through production contrasts with the traditional push production process, driven by a schedule that pushes inventory into stocking locations that may not reflect customer's requirements.

9.17 IMPLEMENTATION OF CONTINUOUS FLOW PRODUCTION (CFP)

CFP is the operational strategy that needs to be carefully reviewed for applicability for a particular production process before implementing any changes. Manufacturing management must be brought into the decision cycle so that a total understanding of continuous flow distribution and manufacturing can be understood.

There are six main steps identified that have to be done before implementing flow strategy:

1. Decide which products or product families will go into your cells, and determine the type of cell: product-focused or group technology (mixed model). For product-focused cells to work correctly, demand needs to be high enough for an individual product. For mixed model or group technology cells to work, changeover times must be kept short.

2. Determine the work elements and time required for making one piece. In full detail, document all of the actual work that goes into making one unit. Time each element separately several times and use the lowest repeatable time. Do not include wasteful elements such as walking and waiting time.
3. Determine whether equipment can meet production capacity. Use a spreadsheet to determine whether each piece of equipment that will be required by the manufacturing cell is set up and capable of meeting the required manufacturing capacity.
4. Create cells' layout. More than likely, more than one employee is working in a cell; however, cells should be arranged so that one employee can do the job. This will ensure that the least possible space is consumed. Less space translates into less walking, movement of parts, and waste.
5. Balance between the cells in the production line. This involves determining how many employees are needed to meet production capacity.
6. Implement cultural shift from "my department" to "my company" as the focus. This global approach to manufacturing will require that all departments participate in the decision process to determine total costs of distribution.

9.18 BENEFITS OF CONTINUOUS FLOW PRODUCTION

When Hamilton Standard Electronic Manufacturing Center (HSEMC) introduced continuous flow production (CFP) into its production line, it eliminated many problems, such as a push system versus pull system. Production bottlenecks were also hidden by the large amount of work-in-process (WIP), and inventory levels far exceeded customer demands. Substantial rework was performed, products traveled long distances in the process, many processes were non-value added, and the cycle time through the process was excessive.

The most important step in Hamilton Standard's CFP improvement process was to obtain the total involvement of all associates from the director level to assembly associates. Training in CFP techniques is provided to everyone at the electronic manufacturing facility.

With a JIT system, problems are not masked by large quantities of WIP, and the root cause of problems has to be addressed and corrected immediately. Employees are certified to inspect their own work with periodic audits by quality to ensure that quality standards are being maintained. Problems are now identified as to the root cause and solved rather than band-aided.

The CFP system is indeed beneficial if implemented correctly and sustained. Continuous flow production is essential for those manufacturers that will compete in the marketplace of the future. The total costs of distribution will be understood and departmental optimization will be eliminated in favor of a total cost approach. Engineering group leaders and managers within manufacturing organizations need to understand, communicate, and implement the vision to streamline production and distribution, establish manufacturing networks, implement manufacturing excellence, and use team-based continuous improvement.

10 Types of Waste

10.1 INTRODUCTION

Waste is any activity that does not provide value to the product. A worker walking long distances across the production line to move a credit application to the next stage of the process wastes time. Unnecessarily, large buffer stocks waste working capital. Production bottlenecks that idle downstream workers waste human resources. Above all, waste wastes money. Customers don't want to pay for your waste, which increases the cost of the product.

Everywhere work is occurring, waste can be found. Sending an installation team to an installation site twice to install a high-speed line because the information needed at a customer premise was incorrect or incomplete is wasteful. So is rejecting a customer request for installation because the application had errors in it that incorrectly indicated that this service was not available at that location. Cost per install goes up and revenue goes down.

10.2 TYPES OF WASTE

In manufacturing, there is always waste, which eventually brings down value and efficiency of production. The following are types of waste in many production lines. Elimination of these types will reduce the cost of products.

FIGURE 10.1 Types of waste.

1. Waste of inventory
2. Waste of overproduction
3. Waste of defects
4. Waste of inefficient process
5. Waste of waiting
6. Waste of materials transportation/movement
7. Waste of motion
8. Waste of rejects
9. Waste of person
10. Other types of waste, depending on the products

In order to understand the waste within manufacturing processes, waste is broken up into elements, as shown in Figure 10.1.

10.3 WASTE OF INVENTORY

Inventory is a drain on an organization, adversely affecting cash flow and often covering poor processes. In a simple way, waste of inventory means storing more parts than the absolute minimum. Waste of inventory increases overhead and hides quality issues in finished goods or work-in-process. It also means having unnecessarily high levels of raw materials, works-in-progress, and finished products.

Extra inventory leads to higher inventory financing costs, higher storage costs, and higher defect rates. Of course, in order to remain responsive to the customer's requirements and ensure control of variance, minimum inventory levels must be maintained. Excess inventory, however, disguises issues like unacceptable changeover times, downtime, and operator inefficiency because there is no sense of urgency to produce product since there is plenty available in storage. Costs escalate when excess inventory is built to accommodate problems in processes. Required safety levels are driven by downtime, quality problems, supplier delivery problems, and job imbalances. In any production line, inventory procedure and inventory storage lists are required and maintained regularly.

Inventory waste is closely connected with waste from overproduction. That is, the overproduction creates excess inventory, which requires a list of extras, including handling, space, interest charges, people, and paperwork. Because of the often substantial cost associated with extra inventory, rigorous measures should be taken to reduce inventory levels:

1. Disposal of obsolete materials.
2. Production only of the number of items required by the subsequent processes.
3. Purchase of required amounts of materials. Cost savings can be achieved through volume orders, and discounts must be carefully weighed against inventory and storage costs.
4. Manufacture of products in required size lots. Measure setup and changeover costs against inventory-carrying costs to achieve the most appropriate size.

It is important to understand that in many operations, inventory covers uncountable other problems. As inventory levels are reduced, these problems will surface, and they must be corrected before inventory levels can be reduced to their optimum levels:

1. Poor scheduling
2. Machine breakdown
3. Quality problems
4. Long transportation time of raw materials or finished goods
5. Vendor delivery times
6. Line imbalance
7. Lengthy setup time
8. Absenteeism
9. Lack of housekeeping
10. Lack of company organization
11. Communication problems within the organization, with suppliers, and with customers

10.4 WASTE OF OVERPRODUCTION

Waste from overproduction is one of the greatest wastes commonly found in manufacturing operations. It is created by producing more products than are required by the market or producing them before they are needed by the market. Both types of overproduction increase the risk of obsolescence, the risk of producing the wrong thing, and the possibility of having to sell the excess items at a discount rate or even discarding or recycling them.

Overproduction can occur because people involved in a process inflate their numbers to cover potential need and because processes are designed with an anticipation of a certain rate of failure. The waste of correction is often amplified by overproduction. Not only do you generate the mistake once, but unknowingly generate the same error multiple times. When the market is strong, this waste may not be very noticeable. However, when demand reduces, the overproduction creates a very serious problem with unsold inventory and all the by-products associated with it:

1. Extra inventory
2. Extra handling
3. Extra space
4. Extra interest charges
5. Extra machinery and equipment
6. Extra defects
7. Extra overhead costs
8. Extra employees
9. Extra services
10. Extra paperwork
11. Extra recycling

Overproduction usually begins by getting ahead of the work required. More raw materials are consumed and wages paid than necessary, resulting in extra inventory. This situation requires additional material handling, storage space, services, and interest paid on money used to carry the inventory. Additional staff, computers, and equipment may be needed to monitor the extra goods. But as serious as these problems are, even more critical is the confusion about what the priorities are needed. People are distracted and unable to focus on immediate goals, which results in additional production control staff. Since the overproduction causes the machinery and operators to seem busy, additional equipment may be purchased and labor hired, under the assumption that they are necessary.

Since overproduction creates difficulties that often obscure more fundamental problems, it is considered one of the most serious types of waste and should be eliminated as promptly as possible. The elimination lies in the understanding that machines and operators do not have to be fully utilized to be cost-efficient, as long as market demands are met. Unfortunately, this concept is difficult for many people to understand. It is helpful for the operator at each stage of production to think of the next stage of the process as his or her customer. Only the amount required by this customer should be produced, meeting the requirements of high quality, lowest cost, and correct timing of manufacturing and delivery.

10.5 WASTE OF DEFECT

The quality of the product or service can be only as good as the quality of the worst component. The waste in a process is magnified by the cost associated with finding and correcting the error as well as the costs incurred by delaying further output while waiting for the correction to be made. Defects relate to both quality and time. In addition to physical defects, which directly add to the costs of goods sold, defects may include errors in paperwork, provision of incorrect information about the product, late delivery, production to incorrect specifications, use of too much raw material, or generation of unnecessary scrap. There is a possibility that defective products can reach customers.

Waste from product defects is not simply those items rejected by quality control before shipment; it actually causes other types of waste throughout the entire manufacturing process:

1. Waiting time is increased in subsequent processes, increasing costs and lead times.
2. Rework may be required to make the part usable, increasing labor costs.
3. Additional labor may be required for disassembly and reassembly.
4. Additional materials may be needed for replacement parts.
5. Sorting the defective product from acceptable parts requires additional labor.
6. Scrapping the defective pieces wastes both the materials and the work already added.

All of the above are serious problems to production. When the customers discover defects in the supply, the company loses the market. Not only are extra warranty and delivery costs incurred, but also customer dissatisfaction may result in loss of future business and market share.

To eliminate product defect waste, a system must be developed to identify the defects. Preventive measures such as quality control procedures must be implemented in all the manufacturing steps. These procedures will prevent any mistakes, caused by humans or machines. Without these preventive measures in place, other time-saving efforts are fruitless.

10.6 WASTE OF INEFFICIENT PROCESS

This waste occurs through processing, such as unnecessary levels of approval for a purchase requisition that provides no value to the product or service. Of all the types of waste, this is often the most difficult to identify. Examples include typing memos that could be handwritten, painting fixtures that are internal to equipment, making color copies when black and white would do, and performing machining work that goes beyond expressed specifications. Waste of human and material resources costs money.

Excessive detailed procedures may be another source of waste. If fixtures and machinery are well maintained, they may require less labor on the part of the operator to produce a quality product. Regular preventative maintenance may also reduce defective product. When the principles of design for manufacture (DFM) are employed and manufacturability is taken into consideration in product design, processing waste can be reduced or eliminated before production even begins.

10.7 WASTE OF WAITING

Waste from waiting is like any other wastes in a production. Waste of waiting is usually readily identifiable by skilled employees who have completed the required amount of work, or employees who spend much time operating machines but are powerless to prevent problems. Waiting for supply materials and waiting for technical service support are two examples of the waste of waiting and are easy to spot.

By completing only the amount of work required, the capacity, both speed and volume of each workstation, can be monitored. This will result in using only the machinery and personnel required for the minimum amount of time to meet production demands, thereby reducing waiting time.

The use of multiprocess employees goes a long way toward eliminating this waste; however, there is always room for improvement. Waiting for finished parts from a previous process, waiting for a document from a document department, waiting for supplies to be delivered, and waiting for a schedule meeting are all examples of waste. Waiting is 100% waste, and whenever it is eliminated the cost is reduced by the entire amount of the wait time. Another principle can be taken in account: first come, first served.

10.8 WASTE OF MATERIALS TRANSPORTATION/MOVEMENT

This kind of waste occurs because of unnecessary movement of materials between production line sections. The transportation and double or triple handling of raw materials/parts and finished goods are commonly observed wastes in many factories. The reason for this type of waste could be a poorly conceived layout of the factory floor and storage facilities, which can mean long-distance transportation and overhandling of materials and semifinished product. Transportation includes any movement of materials or personals, including from one workstation to another. Unnecessary movement does not add any direct value to the product. Therefore, it is vital to convey material and information only when and where it is needed. To preserve profit margins, there should be minimal transportation of parts. In order to minimize transportation waste, improvements must be made in the areas of layout, process coordination, methods of transportation, housekeeping and general organization, or the operation.

10.9 WASTE OF MOTION

As explained above, any movement of people, parts, or equipment that does not add value to the product is waste of motion. Waste of motion can be found in the machine, the method of the production system, and unnecessary personal movement. It is important to recognize the difference between value-added and non-value-added motion when identifying waste in a process or in human motion. There are many times when non-value-added motion can be combined with another part of the process to reduce the overall cycle time. Therefore, minimizing the non-value-added motion will reduce time consumption and eventually reduce costs. Making unnecessary phone calls and walking to get files, documents, supplies, and make copies is

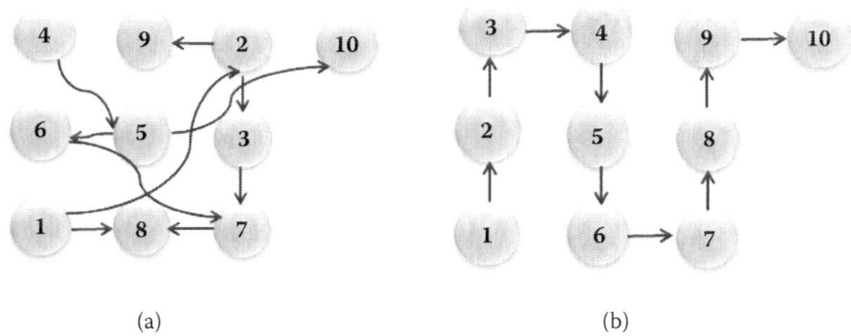

FIGURE 10.2 Waste of motion can be reduced with the right production line layout.

also waste of motion. Specially, these motions are made by skilled employees. Leave these motions to low-skill employees to do. In most cases, it is very difficult to recognize the waste of motion in big companies. It is the most difficult part of reducing this type of waste.

As mentioned above, this type of waste is most often revealed in the actions of the factory employees, as shown in Figure 10.2(a). It is clearly evident in searching for tools, pick and place of tools, and parts kept out of immediate reach of the workstation, and especially by the walking done by one operator responsible for several machines. All of these can be eliminated by carefully planned layout and fixture selections, as shown in Figure 10.2(b).

10.10 WASTE OF REJECTS

Rejects are almost unavoidable in the industry today simply because of the style of production. Error is mostly present depending on the style of production; either in lean manufacturing or in conventional manufacturing style, there is always an error margin to deal with. Rejects give rise to wasted time reworking on the rejects to make them acceptable or losing all the time dedicated to the product all together.

10.11 WASTE OF PERSON

Waste of person is caused by inefficient use of employees' time to work on a specific job. Hiring unskilled employees will lead to wasted time to complete a job without training. The sitting and movement of an employee should be calculated to suit the manufacturing process. This will reduce the manufacturing and handling time, and consequently reduce the cost. This also provides comfortable working conditions to the employee.

As mentioned in the above types of waste, uncomfortable sitting and unnecessary movement of an employee in his or her working cell or the production line will cost money. It also subjects the employee to a serious body fatigue situation. All of these can be eliminated by a carefully planned floor layout of the production line and clearly written manufacturing processes in conjunction with a health and safety committee.

10.12 OTHER TYPES OF WASTE

There are many common types of waste in any small or big manufacturing facility. The following examples are the types of waste from a telecommunication factory: solid wastes generated from soldering processes include component wire trimmings, circuit board trim materials, plastic component reels, plastic integrated circuit packing tubes, scrap electrical components and circuit boards, fiber optic butts and jackets, rejected parts, cleanup rags, general office waste, white and colored paper, different types of packing materials, scrap labels, and label backing paper.

Plants and factories that have cafeterias or food service in break areas also generate food wastes, glass and plastic containers, and other wastes typically associated with serving food. Food wastes are a large percentage of the total waste by weight at plants that have cafeterias.

10.13 WASTE REDUCTION TECHNIQUES

As explained above, there are many techniques to eliminate wastes. However, not in every case can the waste be eliminated. But for many cases, waste reduction should be possible and conducted. To do a waste reduction in a manufacturing line, the following can be performed.

10.14 WASTE OF PRODUCTS

To understand how to reduce wastes in production facilities, it is necessary to know the manufacturing steps. Manufacturing waste reduction generally follows these steps:

1. Parts preparation due to customer's order.
2. Manufacturing process in each section of the production line.
3. Quality control process in each section of the production line.
4. Products are tested and rejects are separated. Most manufacturers have a very low reject rate (1 to 2%).
5. Packaging and shipping products to customers.
6. Customers' feedback and correction and necessary actions should be implemented.
7. Financial analysis must be in place.

10.15 ELIMINATING WASTE

There are many ways to eliminate the waste, as follows:

1. Simplify:
   a. Check product and process design.
   b. Use visual controls (data collection, analysis, data display, etc.).
   c. Apply mistake-proofing in a production line.
   d. Plan employee training.

2. Combine:
   a. One employee can do more than one job at a time.
   b. Reduce number of priceless activities.
3. Eliminate:
   a. Keep standard adjustments as a high product standardization.
   b. Reduce handling time for nonconfirmed materials in a production line section.
   c. Eliminate any types of waste.
4. Verify.
5. Collect data.
6. Analyze data and apply lessons learned.
7. Make modifications wherever important and necessary.
8. Check results with standards.
9. Calculate gain and loss.
10. Document any changes for future references.

10.16 WASTE REDUCTION

Waste reduction includes waste prevention, material reuse, recycling, composting practices, and buying products with recycled content.

10.16.1 Product Design and Manufacturing

Product design is the process of creating a new product to be sold by a business in the market. It is essentially the efficient and effective development and generation of ideas through a process that leads to new products.

In a systematic approach, product designers theorize and evaluate ideas, turning them into possible inventions and products to be used. The product designer's role is to combine design concept and technology to create new products that other people and companies can use.

Old or new product design is not an easy task to put out for production. The product should achieve profit when using efficient manufacturing processes utilizing all the waste reduction techniques. The investors and manufacturers are concerned with production cost. In the end, the investors and manufacturers want an economically produced product for fast payback. There are many ways to reduce the cost of produced product, such as:

1. The purchaser looks at price in many aspects.
2. The end user is concerned with usability and functionality of the final product.
3. The maintenance and repair department focuses on how well the final product can be maintained.
4. Investors and manufacturers want their product to meet customer demand and fast payback.

Manufacturers often make design changes, which may be costly sometimes, but they often cut overall production costs per unit by as much as 30 to 50% due to fewer materials, faster assembly, and using automation. These days, consumers respond very well to green and energy-saving products. In new product design and manufacturing techniques, the reduction of waste can be accomplished in the following ways:

1. Using automatic manufacturing processes to minimize reject rates and achieve repeatability and standardization of the product.
2. Ordering semi- or finished parts for a production line and other components to as exact specifications as possible to reduce cutoffs.
3. Using in-process quality assurance and controls, such as electronic monitoring and control systems, to help maintain quality and reduce rejects.
4. Using rework and recycle processes to reduce the rejects that end up in the garbage.

10.16.2 Materials Reuse and Recycling

Remember the wisdom of give usable and reject products second lives in the production line for a salvaged and good product. Rework inoperable products and reuse subparts in the manufacturing process to produce new products. For example, reuse wire and shrink-wrap spools on site or return them to the vendor. Recycling materials from production processes, such as steel, copper, polystyrene, PVC packaging, scrap equipment, and others, is a good practice in saving materials and the environment. Many manufacturers have implemented cost-effective methods to segregate these materials in production lines. Use different bins in all areas where materials are generated. Bins are commercially available from many different vendors. The bins should be colorful, easy to use, and clearly marked to indicate the appropriate materials to be disposed in each. Reusing materials and recycling make sense economically, environmentally, and socially.

10.16.3 Packaging Reduction

Remember the wisdom of "new packaging concepts = sources of cost reduction" in the production line for packaging products. Cost of packaging processes is one of the essential factors in the total product cost calculations. To reduce the cost, it is very important to review all packaging procedures for adopting new packaging technologies for each product. It is also possible to use reusable packaging whenever possible and accepted by customers. Reusable packaging can be used between production line sections effectively. Several large manufacturers have established closed-loop distribution systems to reduce packaging, labor, and waste disposal costs. Customers also want to reduce waste disposal and recycle costs.

10.16.4 Purchasing

Purchasing raw materials and services for production is one of the important elements in successful companies. Always find alternative materials, and reliable suppliers are very important to update the product quality and cost reduction. Mass purchase materials will reduce cost, but it sometimes depends on the suppliers, materials, and market competition for a particular product.

When a reliable supplier is selected to deliver an item in large quantities, the following are required:

- Qualified supplier
- Approval of supply item
- Monitoring of supplier performance
- Data and technical specifications
- Validation of inspection method and test procedure
- Supply and rejection of items resolution
- Annual auditing
- Tax clearance
- ISO certificate

10.16.5 Office Paper Use

Office paper and stationery costs add to the product costs. Two-sided photocopying can be used with copiers that have this capability. Reuse one-sided copies by using the clean side for note pads, distribution lists, and fax cover sheets. Provide the necessary peripherals to maximize the use of online communications inside and outside the company. Regularly, review and maintain documents on computer disks rather than printing out multiple hard copies. It is easy to reuse two-way envelopes.

10.16.6 Cafeterias and Restrooms

Cafeterias and restrooms create limited types of waste, such as organic waste, disposable Styrofoam plates and plastic utensils, cardboard, aluminum cans, and glass bottles. Many cafeterias and restroom facilities have implemented these waste reduction measures and are saving money. The saved money comes from recycling the plastic and glass, and reducing cleaning costs of the catering tools. Educate employees that a recycling program will save some money and keep a healthy environment.

Organic materials can be composted on site for use in gardens or sent off site. Food accounts for a large percentage of wastes (by weight) at manufacturers that have cafeterias and restrooms. Recently many modifications have been made in restrooms where hot-air hand dryers were installed to eliminate paper towels.

10.16.7 Waste Reduction by Training

Employee training (management, administrative staff, and production employees) should be conducted so they can learn the benefits of waste management and reduction in their areas. Regular training should be scheduled in a yearly plan. Enhance communications and training programs to cover waste prevention, reuse, and recycling issues.

Employee training (for management, administrative staff, and production employees) should teach them to identify and segregate all recyclable materials before placing them in solid waste bins or the solid waste dumpster. Integrate recycling training in new employee orientations, health and safety training courses, and department meetings.

Periodic assessments can be conducted at the facility to identify new waste reduction opportunities and evaluate current measures to reduce waste. Evaluate the operations and waste generation rates during each work shift. Create an information center for management, employees, vendors, and customers that tracks waste reduction efforts, successes, challenges, and newly established targets and goals, for now and in the future.

10.17 ENHANCE YOUR WASTE REDUCTION PROGRAM

As explained above, there are many ways to reduce wastes in a production line. These ways can be practiced in everyday activities. A written waste reduction policy can be established to encourage employees to follow waste reduction to reduce facility costs and impacts on the environment. A waste reduction technical team can be established, composed of management and staff. The team identifies specific goals and objectives. Figure 10.3 demonstrates a flowchart to identifying wastes in a production line.

As shown in Figure 10.3, the process of eliminating waste is quite straightforward, although the individual points may involve a wide range of elements. Below are the definitions in this flowchart:

Normal: A process that is proven to give results.
Abnormal: A process that is not proven to give results. Abnormal functions may be regarded normal after having gone through the process of validation.

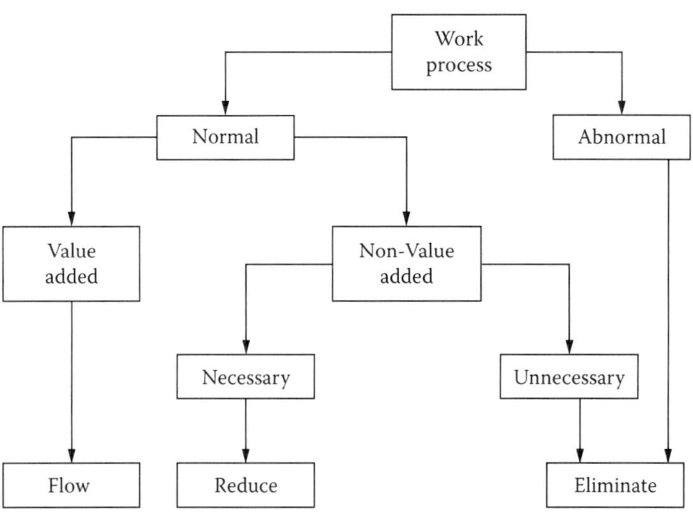

FIGURE 10.3 Flowchart evaluating processes in a production line.

Value added/non-value added: Whether or not value is added to the final product by adding the particular work process.

Necessary/unnecessary: Whether or not the production line needs the particular work process to help produce the products to the wanted standard.

Flow: Add to the flow of the production line.

Reduce: Ensure that the minimum level of the work process is added to the production line so that it does not affect the flow of the production line.

Eliminate: Removed completely from production line.

10.18 REMARKS

Waste does not do any good to manufacturers or customers. Waste elimination is one of the most effective ways to increase profitability in manufacturing and distribution businesses. In order to eliminate waste, it is important to understand exactly what waste is and where it exists in the factory, warehouse, cafeteria, and restrooms. While products differ in each factory, the typical wastes found in manufacturing environments are quite similar.

The first step to eliminating deadly wastes is to identify each one within the operation. After that, measures can be taken to correct the situation and eliminate the problems. Such action may require simple, inexpensive solutions to a single workstation, or may involve changes as massive as a new layout of the factory floor with more efficient machinery. The appropriate solutions require careful study of the operation, clearly defined objectives, and thorough investigation of the benefits to be gained by each change.

11 Application of the 5Ss in Manufacturing Lines

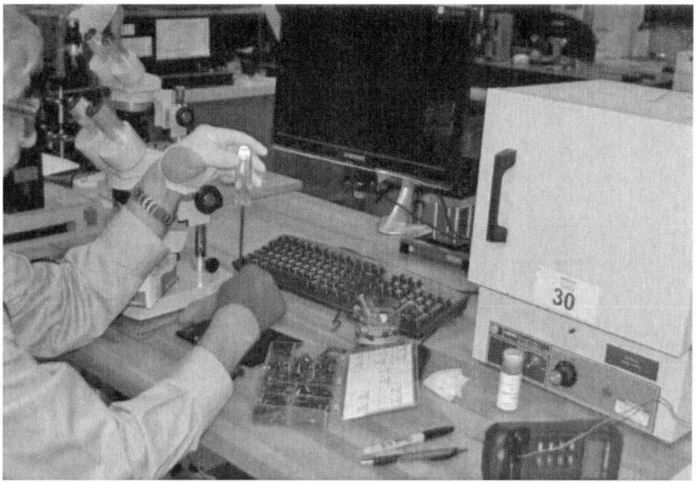

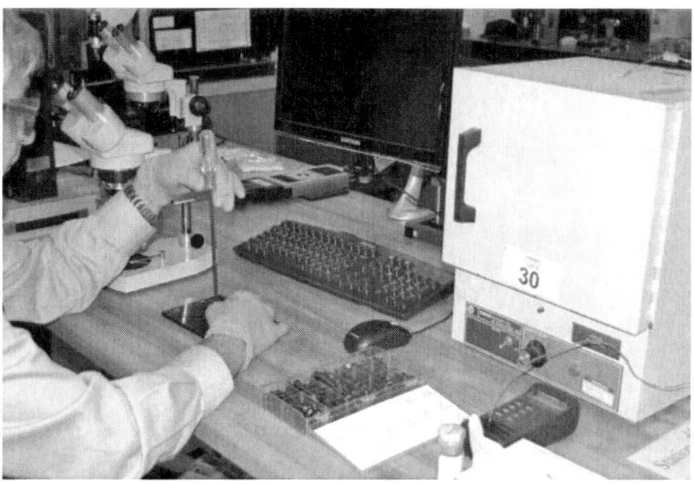

11.1 INTRODUCTION

In order for a manufacturing line to run smoothly, some order must be maintained. Without instruction, a manufacturing line could get out of control and the company could lose valuable time and money. With a 5S system in place, some instruction is kept and the manufacturing line is regulated. This ensures that everything is done logically and without waste, therefore saving time and consequently saving some money.

11.2 THE 5Ss

5S stands for five steps that can be used to reorganize a manufacturing line. They originated from Japan where a high degree of order and discipline were needed. The Japanese names for these steps are:

1. *Seiri*
2. *Seiton*
3. *Seiso*
4. *Seiketsu*
5. *Shitsuke*

In English, their counterparts are:

1. Sort
2. Set in order
3. Shine
4. Standardize
5. Sustain

The 5S's can be explained in detail as follows.

11.2.1 Sort

This is the first S in the system and focuses on the reduction of unnecessary objects in the production line. Any item that is unnecessary for the production line is usually marked with a red tag so that others can see it and take care of the problem, as shown in Figure 11.1. If an item belongs in the production line but is in the wrong area, a red tag is used to identify it. Once all items have been rearranged or removed from the production line, there should be much more available space, which can then be used for more important things, as shown in Figure 11.2. This step also helps with the just-in-case (JIC) mentality.

11.2.2 Set in Order

This step focuses on finding more efficient places to put items in a production line or removing them for storage. For example, it is time-consuming to have someone walk all the way across a room to get to a specific item. Why not have the item he or she needs right beside his or her workstation. Doing this would save quite a bit of time. When performing this step, one should always prepare himself or herself for the job, the time it takes, and how many times he or she repeats the job per a day. Once the employee does this, he or she should be prepared for the job, bearing in mind the job

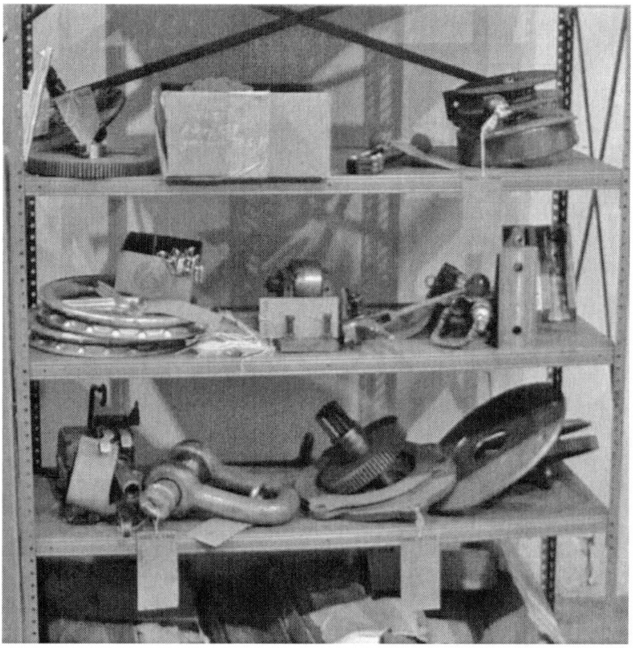

FIGURE 11.1 Stored items with red tags.

(a) Before (b) After

FIGURE 11.2 Sorted items.

should be achieved in a perfect professional way according to the standard procedure for the product.

Every employee should ask himself or herself the following questions:

- What do I need to do my job?
- Where should the items I need be?
- How many items do I really need?

Once these questions are answered, a more effective way of setting up a production line is in order. After completing the job, each item should be cleaned and put back in its place. These items will be ready for anyone to use it with high efficiency.

11.2.3 Shine

Once everything has been put away and rearranged, it is now time to clean. This step requires employees to actually clean the production line. This is done by sweeping, mopping, and all other forms of cleaning. It is recommended to perform this step fairly often since a clean line does have a positive effect on employees. Employees can make a production line their own in a way and take pride in cleaning it. It will also allow for a faster recognition of problems in the production line, such as oil leaks, spills, or water use in the area. If an area is kept clean, problems like that can easily be spotted. It also keeps the equipment working well; if it is cleaned regularly, employees can quickly identify problems with a machine before is too late.

11.2.4 Standardize

This is the idea of setting rules or guidelines that can create standard practice in the workplace, and also involves training employees to do jobs correctly. This allows an employer to get the best work out of his or her employees, and will also allow employees to maintain the previous three steps (sort, set in order, and shine) on their own. Having standardization can allow an employer to give his or her employees some freedom, knowing that the job will still get done. Applying standardization in each production step produces a standard product regularly.

11.2.5 Sustain

This step requires that employees continue to perform the 5S's in the workplace and not allow the workplace to get dirty and cluttered. This can be difficult because most employees feel more comfortable going back to old habits. Because of this, the workplace has a higher chance of going back to the way it was before. To help implement the 5S's in the workplace, internal groups should be made. This would ensure that the workplace sustains the clean tidy look that employers want to achieve.

All of these steps can greatly help a production line. Some of the good things that can come out of implementing the 5S system in a workplace are:

1. More profit
2. Higher-quality products
3. Improved efficiency
4. Improved safety
5. Reduced space requirements
6. Reduced machine downtime
7. Higher employee morale
8. Saving some money
9. Comply with International Organization for Standardization (ISO) standards

These are all great benefits, but without the proper training and time, the rewards won't be as high. Therefore, in order for the best possible outcome, one must properly train his or her employees the basics of the 5S's. Once this is complete, the outcome may be more profitable. Moreover, an information center can be created to show the deference before and after applying the 5S's. Sometimes a personal award can be presented to employees who spend recognized time and energy to achieve the goal of 5S's in the production line.

11.3 IMPLEMENTATION OF THE 5Ss IN A MANUFACTURING LINE

The following steps are taken into consideration in preparing for 5S implementation.

11.3.1 Plan

An action plan can be prepared before the beginning of any action.

1. Place (room, production line name/number, lab, storage, etc.)
2. Date (time, day, month, year)
3. Estimated time (how many hours)
4. Create group(s) and assign responsibilities
5. Internal and external recourses needed (management representative, manufacturing engineering group leader/manager, process, test, quality control (QC), auditing engineers, building maintenance, electrician, plumber technicians, health and safety officer)
6. Clear instruction
7. Hotline (phone, pager, e-mail, etc.).
8. Plan put together
9. Plan approved

11.3.2 Start

Once the plan is finalized and approved by the manager, the action can be started immediately. This is outlined in the following steps:

1. Take pictures for current situations of the production line (stations, employees, ovens, scaffoldings, cabinets, etc.).
2. Take pictures for employee locations and sitting styles.
3. Prepare a list of breakdown and out-of-calibration tools, devices, instruments, equipment, etc.
4. Identify the most frequent failed devices, instruments, and tools.
5. Prepare copies of the updated process.
6. Prepare a checklist.
7. Prepare the resources:
   a. Required new tools
   b. Required bins (manufacturing, handling, storage, incoming materials, outgoing products, dustbins, etc.)

c. Spare parts and consumables
d. Labels (safety labels and signs)

11.3.3 Put Plan into Action

The plan may be divided into two actions, depending on the production facilities, as follows:

11.3.3.1 Light Actions (Simple Jobs)

Light actions can be carried out by skilled and unskilled employees under the supervision of a supervisor or engineer. These are actions such as:

- Rearrange labels.
- Rearrange bins.
- Update process folders.
- Use red tags to get rid of unwanted or personal items from workstations, production line, and clean room.
- Lay out wire connections and power cables.
- Lay out air hoses.
- Replace burnt-out bulbs and fluorescent tubes.
- Replace worn-out hand tools.
- Replace worn-out instruments, jigs, etc.
- Rearrange swab containers.
- Check chemical containers.
- Check and replace the designated red bins for scrap blades and fiber cable pieces left over from the cleaving process.
- Check first-aid and eye-wash stations.
- Update safety signs ("You Are Here," emergency, fire distinguishers, etc.).
- Check the Workplace Hazardous Materials Information System (WHMIS) control handbook, which should be available in designated locations.
- Update the material safety data (MSDS) folder, which should be available in a designated location.
- Check and clean epoxy containers and freezers.
- Tidy up the change room, lockers, smoke, and head and foot wares cabinets.
- Clean station surfaces and floors.
- Take pictures during and after the compilation of the work in this plan.

11.3.3.2 Heavy Actions (Complicated Jobs)

Heavy actions can be carried out by skilled employees under the supervision of a supervisor or engineer. These are defined as:

- Relocating stations, tables, ovens, cabinets, scaffoldings, etc.
- Calibrating instruments, tools, jigs, air supply pressure, temperature controls, micrometers, hot plates, scales, oven temperature, etc.

- Updating computers, test systems, and data collection and analysis systems software
- Checking workstation of light intensity level
- Checking production line room of light intensity level
- Checking the lens filters and light intensity of the microscopes
- Maintaining and calibrating automated machines
- Maintaining and calibrating epoxy containers and freezers
- Taking pictures during and after the compilation of the work in this plan

11.3.4 Analyses

After any action plan, data collection and observation can be analyzed and lessons learned from:

- Measuring results
- Verifying changes
- Presenting results and highlighting advantages and disadvantages
- Organizing a seminar to present achievements and discuss new future ideas
- Celebrating and rewarding
- Creating an information center and displaying the achievements and future plans

11.3.5 Future Plan

For the future, action plans must prepare to keep the 5S's in daily practice for every employee in the manufacturing facilities. The following steps are:

- Creating preventative maintenance plan
- Creating schedule maintenance plan
- Creating employee training plan
- Creating follow-up plan
- Making 5S's a part of daily work practice
- Creating a hotline for new ideas

12 Management and Employee Responsibilities

12.1 INTRODUCTION

Each company has a different management structure. Companies are going to create and obey their own rules and regulations in order to manage their business. Any company can be chosen as an example of the management involved in the formation of a high-tech company; as long as the company stays in the competitive market, this company can be successful in its business.

12.2 STRUCTURAL NEEDS OF A COMPANY

The structure of any manufacturing company is split up into departments and special groups. The departments and special groups could be technical or managerial. The first and highest ranking is the president of the company or the chief executive officer (CEO). This position oversees the entire company, makes all the big decisions, and ensures that all company locations meet expectations. The president of the company has an executive group to support him or her in management decisions. The president and executive group are supported by a secretarial office. The executive management also deals with issues on a global level, perhaps between multiple facilities and other companies in the same field.

Normally, each company has a human resources department that is responsible for managing employees' resources, customer advocacy, and information technology. The financial department covers accounting and finances of the company.

The technical departments are for manufacturing processes and their names depend on the products. The technical departments are responsible for overseeing the process/manufacturing, engineering, sales, marketing, and science and updating technology. Big companies have a special group called research and development (R&D). This group is responsible for solving manufacturing problems, modifying products, and implementing new products. Testing and measurement departments are self-explanatory. They manage the testing and measuring stages for each product. Each manufacturing company needs a packaging, shipping, and receiving department that is in direct contact with customers. The sales and marketing departments deal primarily with sales and marketing operations of the business. Technical, sales, and marketing departments coordinate the final sales and find new markets for the products. The sales and marketing departments inform the company of new orders from current and new customers.

12.3 RESPONSIBILITIES OF EMPLOYEES

The information concerning the employees and their responsibilities completely depends on the location and type of employment. However, here are some of the general forms of responsibilities that can apply to a variety of different organizations:

- Fulfilling the duties and responsibilities established in one's job description
- Meeting applicable performance standards
- Performing tasks carefully
- Following supervisors' instructions
- Attending education and training programs that are needed to increase awareness and understanding of one's job position
- Taking all the necessary steps to safeguard assets against waste, loss, unauthorized use, and misappropriation
- Reporting breakdowns in internal control systems to the supervisor
- Keeping and protecting professional/business secrets by keeping them confidential
- Observing the benefits of being an employer
- Refraining from the use of one's official position to secure unwarranted privileges
- Coordinating and cooperating with all other employees to try to eliminate workplace injuries and illnesses
- Applying the principles of accident prevention in daily work and using proper safety devices and protective equipment as required by the employment or employer
- Not wearing torn or loose clothing while working around machinery
- Reporting promptly to the supervisor every industrial injury or occupational illness
- Doing everything necessary to protect the life and safety of employees within reason
- Applying the Workplace Hazardous Materials Information System (WHMIS), material safety data sheets (MSDSs), and electrostatic discharges (ESDs) in the workplace whenever necessary to protect employees and oneself

12.4 EMPLOYEE RIGHTS

One of the most important highlights for an employee is that every employee has rights. Some of the main employee rights can be described as:

- To a specific job position and job description according to the collective labor agreement or the collective bargaining contract
- To receive a salary that is in accordance with the collective labor agreement or the collective bargaining contract
- To protection; this is provided by law and legal agreements
- To belong to a union
- To a healthy and safe working environment according to provincial and federal laws

12.5 RESPONSIBILITIES OF MANAGERS

- Some of the main responsibilities of a manager can be described as:
- Maintaining an office environment that encourages employees to understand the purpose of such policies and procedures as well as support a positive internal control environment
- Documenting policies and procedures that are to be followed in performing office functions
- Identifying the control objectives for the functions and implementing cost-effective controls designed to meet those objectives
- Regularly testing the controls to determine whether they are performing as intended

12.6 SAFETY ETHICS

Safety ethics can be enforced by the following.

12.6.1 Laws and Regulations

Laws and regulations mainly inform individuals to follow them to achieve the job requirements with high standards and safety. Ethics instruct individuals and organizations to do the right activities lined up with the laws, regulations, and standards.

12.6.2 Management and Ethnics

There are many aspects of traditional approaches to safety management. It could be viewed as less than ethical, but safety programs based on "doing the right thing" are better and more sustaining in order to manage safety programs. Management is ultimately responsible for building an ethical system that can effectively examine and control workplace hazards.

Perhaps the greatest economic reason to support an ethics-based approach to safety management within a manufacturing system is that prosperity generates an environment where continuing production improvement and reduced risk can be achieved. To enable management to use an ethical safety management systems approach, safety professionals need to promote a more ethical approach to managing their own profession. This approach requires moral courage, belief, and professional agreement. Management should include all level of employees in any approach.

On a daily practice, employers need to consider the compromise between costs and safety. Workplace safety has implications of expense and time with business decision makers. Managers must understand the importance of an employee's safety, and not consider it as a sunk cost but as an investment for business improvements as well as for better insurance costs.

Any safety-fast approach and strategy that focuses on employees will require moral conviction and professionalism, despite pressures from top management to control increasing costs. Safety programs based on ethics are better for sustainment and the future management of safety issues. Safety plays a role in helping management meet the demands of business strategy, financial improvement, and increasing operational performance.

An established workplace safety plan will produce ethical outcomes such as integrity, open communication between management and employees, and management credibility. The workplace safety plan will drive an organization toward profitability and sustainability. Management must lead a culture of professional practices that value safe behavior for the benefit of ethics and for the purpose of increasing employees' morale and productivity.

12.6.3 Occupational Safety and Health Administration

Provincial and federal safety procedures and legislations help to minimize risk and control hazards. Any violations may bring serious consequences, such as fines and legal consequences, and even loss of a business license. Ethical principles demand that management, employers, and unions do more to protect their employees, and create a culture of constructive daily practices. Managers are ethically obliged to use their authority to create a healthy working environment. However, proper attention to workplace safety can result in improved employee morale, increased job security, and customer satisfaction; this keeps the business healthy in a competitive market.

Employees expect management to act responsibly and to put employee health and safety first by establishing and practicing the safety workplace plan. Employee health and safety should have priority over business profits. In order to create a safe working environment, employers can practice the following principles:

- Set up a safety motivation program to inspire employees to comply with workplace safety guidelines, roles, and regulations.
- Adopt a zero-tolerance policy on workplace violence and safety violations.
- Conduct ongoing safety training for managers, group leaders, and employees.
- Update safety guidelines, roles, and regulations.

- Have easy documentation access to all workplace safety guidelines, roles, and regulations.
- Ensure employees work hand in hand with management to achieve safety goals.

These practices, in the long run, will save time, money, and eliminate workplace injuries. For optimal effectiveness of workplace safety, managers should always take necessary actions to check on employees operating in the workplace. These actions should be followed up with an analysis for the improvement of future plan modification. Companies should have safety programs blended into their business and future plans culture.

A safer workplace has healthier employees, reduces costs, and increases performance. This leads to higher quality and customer satisfaction. Employers must go above and beyond what is required and create a culture of care and respect for their valuable assets and employees in the workplace.

12.7 SOCIETY AND ETHICS

Society and ethics can be banded by the following.

12.7.1 Human Society: Family and Person

A human society is defined as a group of people living in a particular country or region and having shared customs, laws, cultures, ethics, and organizations. The society has responsibility toward its members. Many societies have a tradition of ethics, which transfer from the old to the new generation. These ethics are built on traditional culture and religious fundamentals. The new generation should respect these values and try to maintain them. Each person in the society should practice good values in his or her house and work. Government may change some of these values by introducing new laws.

12.7.2 Professional Societies

There are many professional societies around the world that represent a group of professionals. For example, engineers uphold and advance the integrity, honor, and dignity of the engineering profession by:

- Using their knowledge and skills for the enhancement of human welfare and the environment
- Serving honestly, with fidelity, their employers, clients, and the public
- Striving to increase the competence and prestige of the engineering profession
- Supporting the professional and technical societies of their disciplines
- Supporting new engineers to build up their knowledge of technologies, ethics, standards, codes, and market values

12.8 FUNDAMENTAL STANDARDS IN PROFESSIONAL ENGINEERING SOCIETIES

Professional engineering societies should carry on fundamental standards in their jobs:

- Professional engineers should hold the principles of the safety, health, and welfare of the public and should comply with the principles of sustainable development in the performance of their professional duties.
- Engineers should issue public statements only in an objective and truthful manner.
- Engineers should perform services only in the areas of their proficiency.
- Engineers should act in professional matters for each employer or client as faithful agents or trustees, and should avoid conflicts of interest in daily activities.
- Engineers should build their professional reputation on the merit of their services and shall not compete unfairly with others.
- Engineers should act in such a manner as to uphold and enhance the honor, integrity, and dignity of the engineering profession and should act with zero tolerance for bribery, fraud, and corruption.
- Engineers should continue their professional development throughout their careers, and should provide opportunities for the professional development of those engineers under their supervision.

13 Lean Manufacturing

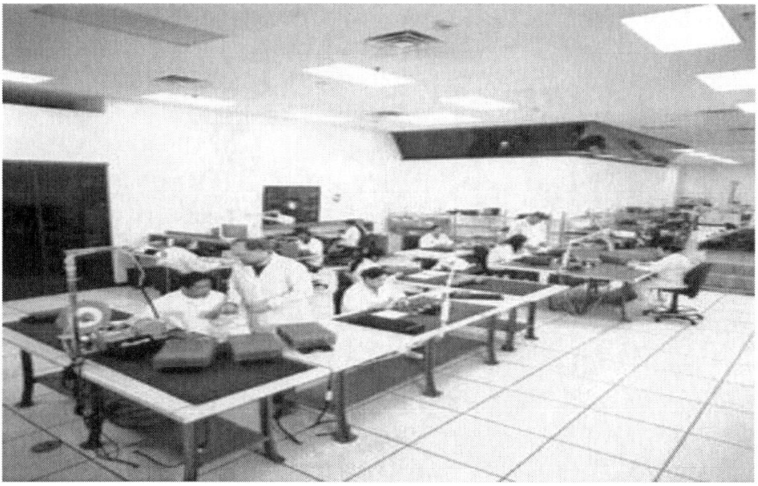

13.1 INTRODUCTION

Lean manufacturing is a philosophy of production that emphasizes the use of the minimum amount of all resources (especially labor and time) used in the various activities of the manufacturing process. Lean manufacturing involves identifying and eliminating non-value-adding activities in design, production, supply chain management, and dealing with customers. Lean manufacturers employ teams of multiskilled employees at all levels of the organization and use highly flexible, increasingly automated machines to produce volumes of products in potentially massive varieties. Lean manufacturing contains a set of principles and practices to reduce cost through the relentless removal of any type of waste and through the simplification of all manufacturing and support processes.

The production line in lean manufacturing is designed as a cell. The employees in the cell are completely responsible for manufacturing an optical device. Each employee is doing more than one job. The production line has a certain number of devices to be manufactured by the cell during the day. If there are more devices required to be produced, then more cells have to be added to the production line. The cell is arranged as a horseshoe or line shape. The shape depends on the manufacturing device steps. On each station, there is a box. The box contains a stack of trays. The trays move from one station to another, and each employee does one job or more according to the manufacturing process.

In summary, lean manufacturing is an industry performance improvement tool that focuses on enhancing quality, cost, delivery, and employees. Employees and managers work together to achieve the goal without any individual holding responsibility for any errors, as it is the responsibility of everyone involved.

13.2 HISTORY OF LEAN MANUFACTURING

Lean manufacturing has evolved in North America from its beginnings in the Toyota Production System (TPS) in Japan. Many of the most recognizable phrases are Japanese terms that have become standard terms in lean manufacturing.

In manufacturing processes, TPS adopts the following:

- Muda: Eliminate anything that is considered waste.
- Kaizen: Improve or change for the better.
- Kanban: *Kan* means "visual," and *ban* means "card." It means a signaling system using cards.
- Poka-yoke: Mistake-proofing. It means no mistakes allowed in the manufacturing process.

13.3 KEY ELEMENTS OF LEAN MANUFACTURING

In general terms, key elements of lean manufacturing work together to continually improve the production process. Put simply, waste elimination is accomplished through just in time and jidoka, maintained through standardized work, and improved through kaizen.

13.3.1 Just-in-Time (JIT)

Just in time is producing what is needed, when it is needed, in exactly the amount needed by using pull systems (i.e., kanban), continuous flow processing, and synchronizing the production speed.

13.3.2 Jidoka

The ability of production to be stopped in the eventuality of a problem, by either the machines themselves or people using "stop systems" and error-proofing.

13.3.3 Standardized Work

This is to standardize procedures concentrating on the most efficient human movements and work sequence for each process by using synchronous production speeds, a working sequence, and standard in-process stock. It is never-ending job design through continuous improvement using a basis of standardized work.

In summary, the principles from the history of lean manufacturing are to reduce waste through the application of a number of process improvement tools.

13.3.4 Understanding the Application of Flow Processing Techniques

Flow production is based on the just-in-time methodologies that fall within a lean production system. The concept of making products flow is almost counterintuitive and difficult for many people to conceptualize let alone implement. Flow means producing without (or with minimal) batches, waiting, or wasted motion. In other words, flow means that each individual product keeps moving either in the literal sense or in the sense of being continuously worked on.

13.4 APPLYING LEAN PRINCIPLES (POKA-YOKE)

Poka-yoke is Japanese for "mistake-proofing," and it is a key tool in a proven methodology for achieving remarkable efficiency in manufacturing. Known as lean, this methodology depends on a simple but powerful principle: eliminate waste. Initiating, establishing, and maintaining a lean operation entails these interrelated steps:

1. Identify customer value. Define value from the perspective of the customer. Understand clearly and exactly what product or service the customer desires, when it is to be delivered, and at what price. No matter how valuable a particular aspect of a product or service may appear to you, it is irrelevant if it is not similarly valued by the customer.
2. Document the value stream. Precisely map the set and sequence of all specific actions, communications, and material movement required to bring a product or service valued by the customer from conception to final delivery.
3. Recognize the waste. Mapping the value stream enables you to identify value-adding and non-value-adding activities from the customer's perspective. Any activity that does not add value for the customer is waste and offers an opportunity for improvement.
4. Standardize processes. Determine the one best way to perform a step or a process and then follow it. Until you consistently follow the process steps to provide your customers with what they need, you cannot be sure of sustained improvement. To eliminate variation in processes, you must clearly and in detail define the sequence, timing, and outputs of each activity. Standardized processes also enable you to expand production with minimal disruption.
5. Establish continuous flow. Ensure the uninterrupted movement of material through a process without backflow or scrap, one piece at a time. Continuous flow yields shorter cycle times and shorter lead times. It allows production flexibility, higher throughput, and increased revenue. Remember, it is a value stream—make it flow smoothly.
6. Enforce quality at the source. Make sure that things are done right the first time, eliminating after-the-fact quality inspections that may catch errors often after the mistake has been duplicated over and over. First, the processes should be designed to keep defects from occurring. The strategy is referred to as poka-yoke. Second, when defects do occur, they should be detected and corrected where they occur by workers in the in-line process. Personnel at each stage in the process need to know the quality specification limits, operate within those constraints, and build in controls to avoid poor quality from advancing in the process.
7. Strive for continuous improvement. As lean uncovers waste, often new layers appear that offer additional opportunities for continuous improvement

or kaizen. Every employee in a process should be encouraged to be on the lookout for ways to continuously improve the process and to eliminate non-value-adding waste.

13.5 KEY TOOLS OF LEAN MANUFACTURING

Lean employs a variety of tools to put those principles into practice. Some representative examples include:

1. Process mapping, which may be thought of as a subset of value stream mapping, visually displays precisely how a particular process is carried out. The map reflects what actually happens rather than what you believe should happen so that opportunities for improvement can be uncovered and standardized processes developed.
2. Standard work provides detailed and complete operating procedures for a process or activity and communicates those procedures clearly to the personnel who must perform them. In the absence of detailed, standard work specifications, workers are likely to make assumptions about how to perform work and, as a result, generate waste.
3. Work cells, often laid out in a U-shape, bring together several stages of a process in order to eliminate transport waste and waiting, to facilitate one-piece or small-batch flow of products through the process, and to take advantage of multipurpose workers who can perform any process handled by the cell.
4. 5S (sort, set in order, shine, standardize, sustain) specifies rules for cleaning and organizing the workplace so that each worker's work area is laid out and maintained for maximum efficiency.
5. Level production by distributing volumes and product mix as evenly as possible over time in order to avoid disruptive peaks and valleys.
6. Poka-yoke enables the enforcement of quality at the source by providing methods of mistake-proofing through in-line quality testing of 100% of the units in the process.
7. Kaizen circles and kaizen events increase worker involvement and effectiveness by bringing together small groups of workers to generate ideas for solving problems and improving processes, thus helping fulfill the ongoing goal of continuous improvement.

13.6 DIFFERENCES BETWEEN LEAN AND CONVENTIONAL MANUFACTURING

One of the stumbling blocks to lean manufacturing is the concept and source of waste. In any manufacturing environment, every employee works to minimize waste. Employees learn waste sources and types from work experience and workshops. However, lean manufacturing redefines the sources and types of wastes.

TABLE 13.1
Differences between Conventional and Lean Manufacturing

Conventional Manufacturing	Lean Manufacturing
• Work is fragmented and employees are nonskilled.	• Work is performed by teamwork where possible and employees are multiskilled.
• Employees are limited to a narrow job.	• Employees have limitless duties.
• Doing and thinking are separate.	• Doing and thinking are combined.
• Employees are mainly interested in pay and are easily replaceable.	• Employees want to do a good job and are valuable resources trained in new skills and the adoption new technologies.
• Pay is geared to the job and determined mainly by a job classification system.	• Pay is linked to the skills required; group incentive and profit sharing plans are used to enhance commitment.
• Managers and engineers are the experts and they give orders. See Figure 13.1.	• Employees are experts. There are no supervisors and managers to coordinate their activities. Managers, engineers, and employees together act as coaches. See Figure 13.2.

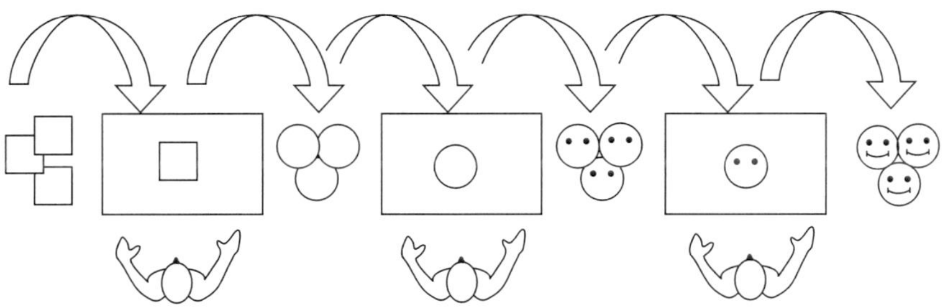

FIGURE 13.1 Conventional manufacturing in a production line.

Traditionally, waste has been viewed as an object. It is very easy to envision a barrel of scrap and identify it as waste. In lean manufacturing, the term *waste* actually refers not to the physical material, but rather to the relationship of the resource to the end customer. In short, if the end customer would not pay you for it, it is waste. Table 13.1 represents some differences between conventional and lean manufacturing in productions.

13.7 LEADERSHIP IN LEAN MANUFACTURING

As explained above, managers and engineers act as coaches. The role of coaches within the organization is the fundamental element of sustaining the progress of lean thinking. A key focus area for coaches is PDCA thinking, which means:

Plan: Establish the objectives and processes necessary to deliver results in accordance with the specifications.

Do: Take actions to implement the lean processes.

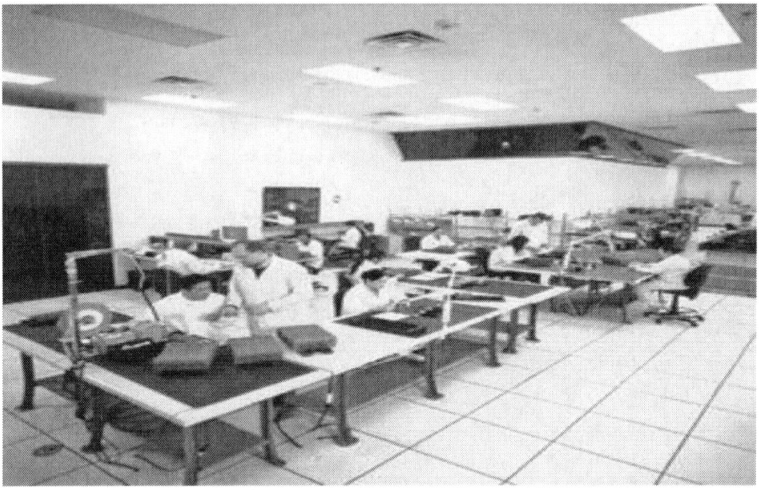

FIGURE 13.2 Diagrammatic representation of lean manufacturing environment.

Check: Monitor and evaluate the processes and results against objectives and specifications and report the outcome.

Act: Apply actions to the outcome for necessary improvement. This means reviewing all steps (plan, do, check, act) and modifying the process to improve it before its next implementation.

13.8 LEAN MANUFACTURING AND WASTE

As explained in this book, lean manufacturing focuses on the reduction of waste types to improve overall customer satisfaction. Lean manufacturing emphasizes the following:

1. Overproduction: Manufacturing more than what is needed by customers or manufacturing too early.
2. Transportation: Moving products farther than is minimally required.
3. Waiting: Products are stalled while waiting on the next production step, or people waiting for work to do.
4. Inventory: Having more inventory than is minimally required. Excess inventory is the deadliest type of waste.
5. Motion: People moving or walking more than minimally required.
6. Processing problems: Stand-alone processes that are not linked to upstream or downstream processes.
7. Defects: The effort involved in inspecting for and fixing defects and reducing rejects.
8. Safety: Unsafe work areas create lost work hours and also cost money.

13.9 LEAN MANUFACTURING IMPLEMENTATION PROGRAM

Let us look at an example of what a lean implementation program would be:

1. Management agrees and discusses its lean manufacturing vision.
2. Management brainstorms in order to identify project leaders and set objectives.
3. Plan and vision are communicated to the employees.
4. Ask employees to form the lean manufacturing implementation team (five to seven employees, all from different sections or departments). The lean manufacturing implementation team may include a manager, an engineer, a supervisor, employees in the production line, a safety officer, production support, finance, and human resources.
5. Appoint members of the lean manufacturing implementation team.
6. Make and put the plan into action.
7. Measure, analyze, and evaluate results and encourage feedbacks.
8. Celebrate successes.
9. Create and schedule future plan.

13.10 CHARACTERISTICS OF LEAN MANUFACTURING

Here some of the lean manufacturing characteristics that can be applied:

1. Continuous improvement through teams.
2. Total quality management is focused in all aspects of the company's operation (zero defects).
3. Implement the pull system to produce only what the customer needs.
4. Employee qualification and training.

13.11 APPLICATION OF LEAN MANUFACTURING

Below are application examples of lean manufacturing principles.

13.11.1 Lean Software Engineering

Lean manufacturing and the Toyota Production System (TPS) concepts have inspired successful company software development methodologies. These software development methodologies apply the lean concept to the architecture, design, and construction/implementation software development activities. In a separate but related effort, lean software development has adapted TPS to the management and measurement of software development projects.

13.11.2 Lean on Healthcare

One of the recent trends in healthcare has been the application of lean principles in improving patient care and reducing medical errors. The Pittsburgh Regional Health Initiative has been one of the leaders in applying lean and TPS methods in hospitals. Dr. Robert Davies has written a book describing how lean Toyota methods may be applied to healthcare. Steve Spear, formerly of Harvard, and now with the Institute

TABLE 13.2
Some Dos and Don'ts

Do	Do Not
• Ensure your strategy passes the test.	• Put stretch targets before learning targets.
• Involve a wide range of employees in measure creation.	• Limit employees' participation.
• Validate a link between measures and desired outcomes.	• Leave the scorecard measures out of the planning process.
• Empower, enable, and encourage.	• Rely on the scorecard for all your information needs.
• Develop systems to produce the scorecards.	• Expect staff to manually prepare scorecards.
• Be good at analyzing trends.	• Ignore data analysis.
• Include measures that encourage innovation and opportunity taking.	• Forget technology development.
• Revalidate the effectiveness of measures annually.	• Forgot to evaluate the plan annually.

for Healthcare Improvement, has written multiple *Harvard Business Review* articles on applying the "Toyota DNA" to healthcare settings.

13.11.3 Lean Manufacturing Strategy

As Dr. Robert Davies has written, what are some dos and don'ts in order to heighten the probability of success? Table 13.2 presents some dos and don'ts in lean manufacturing strategy.

13.12 BENEFITS MAKE LEAN MANUFACTURING SUCCEED

These are some general benefits that make lean manufacturing successful in many manufacturing lines:

1. Improvement plans have a clear focus.
2. Environment is changed to "no blame."
3. Results are timely and visible.
4. Open and full communication.
5. Emphasis on training.
6. Team and individual results are recognized.

13.13 ADVANTAGES OF LEAN MANUFACTURING

These are the advantages of using lean manufacturing techniques in production lines:

1. Productivity improvement.
2. Total manufacturing time is saved.

3. Less equipment utilization (machine time, wear and tear).
4. Reduces scrap and rejects—material cost saved.
5. Low inventory levels—stock holding cost saved.
6. Quality improvement—keep high standards.
7. Plant space saved—with a more efficient layout.
8. Better labor utilization—employees are used to perform more than single tasks.
9. Makes everyone accountable.
10. Train employees to do a multitude of tasks, which gives adaptability and flexibility.
11. Morale improvement in employees.
12. Management plans the work.
13. If lean manufacturing is implemented properly and used in daily activity, there are constant cost savings in production lines.
14. Overall cost reduction.

13.14 DISADVANTAGES OF LEAN MANUFACTURING

Some of the disadvantages of lean manufacturing applications in production lines are:

1. High training cost for the employees
2. Not easy to implement in some cases
3. Hard on some employees and high rate of fatigue failure
4. Employees work like automated machines
5. Cannot be implemented in small production lines
6. Cannot be implemented in every production line
7. Long-term commitment

14 Product Improvement

14.1 INTRODUCTION

Product improvement is the process of modifying and improving the product by examining and making strategic changes to processes in the production line in order to produce a better, more efficient product. Product improvement involves dissecting the product, taking it apart, making it new, and reconstructing it in a different way. This is one of the most important aspects of a production line, because there is always room for improvement of a product. In order for a company to develop and remain competitive, there is always a need for continuous improvement of the products by adopting a new technology, using new materials, advancing the market, and saving costs to suit customer requirements. Figure 14.1 shows a new car model, which has the latest safety, comfort, dynamics, and maneuver advanced technologies.

14.2 PURPOSE OF PRODUCT IMPROVEMENT

There are several reasons for product improvement:

- Increasing the quality of the product
- Increasing safety
- Increasing job security
- Increasing customer service
- Increasing employees' participation

(a) Old model car.

(b) New model car.

FIGURE 14.1 Old and new car models.

14.3 PROCESS OF PRODUCING ECONOMICAL PRODUCTS

The following steps will enable the reduction of a product's cost:

1. Study for the market and for consumer demand.
2. Plan a production line.
3. Create a production line.
4. Create a research and development group.
5. Improve the products regularly.
6. Use alternative and high-quality materials.
7. Train and educate employees, i.e., management and technical employees.
8. Use economical statistical cost analysis and make products competitive.
9. Create short- and long-phase planning.

14.4 MODIFY FLOOR PLAN

There are many ways to modify the floor plan of a production line. The following are the most common.

14.4.1 Floor Plan Design

In manufacturing, the quality of the finished product is always dependent on the processes in the production line. A floor plan is a scaled visual representation of the arrangement or organization of the production line indicating the relationship between space, employees, devices, and the flow of the product. The floor plan is drawn up in order to maximize the space, arrange jigs and devices, and organize employees in order to minimize production, handling, and shipping times.

14.4.2 Modifying a Floor Plan

Modifying a floor plan helps in the utilization of space in the production line, i.e., using the minimum size of work area that the operator requires to complete the job. A floor plan can aid in making calculations as to how many employees, devices, and equipment are required in a specific clean room or production line. The utilization of space saves the company money by saving time due to handling. The application of the 5S's can help in maximizing space by removing unnecessary items.

Figure 14.2 is a conventional production line before any modifications. The figure shows that this production line has not been modified in order to maximize the space. Figure 14.3 is a diagram of the same production line after modification. The same production line has been moved into a smaller space, and the space has been maximized to fit the same amount of stations and employees. This improvement allows space to be utilized for a production line that requires more area.

14.5 IMPROVING AND OPTIMIZING HANDLING AND PROCESS TIMES

Modifying the production line helps in reducing handling and process times in the production line. Modifying the production line also helps reduce the waste that occurs due to equipment repairs, material, and shortages. It also reduces the waste from transportation due to the unnecessary movement of materials or products. When the production line is designed in an effective manner, products move in unison, reducing the processing and handling times.

14.6 REDUCE THE WASTE OF INVENTORY

Reducing raw material waste, subassemblies, and unfinished goods will reduce the production cost.

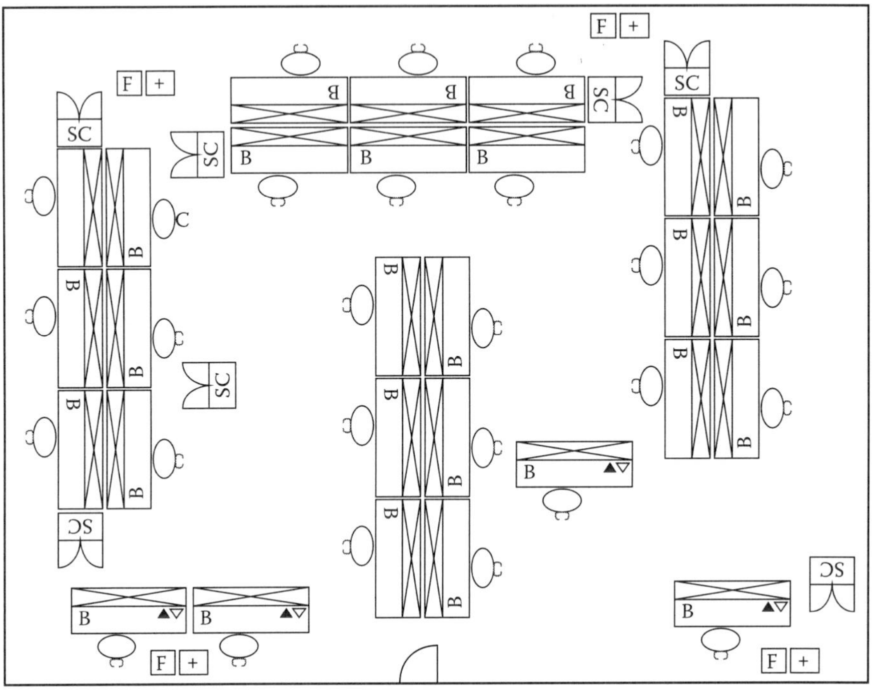

FIGURE 14.2 A conventional production line.

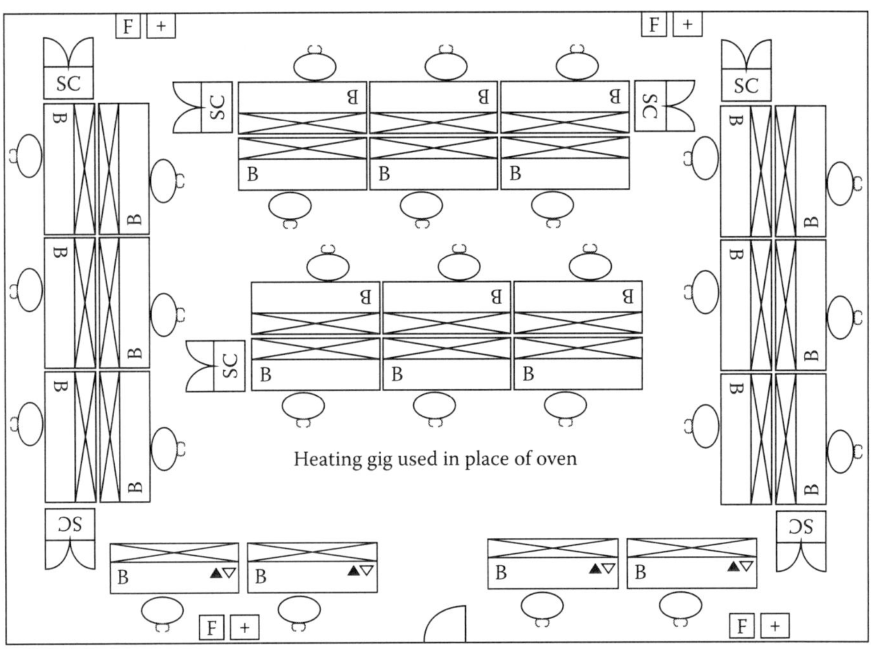

FIGURE 14.3 Modified conventional production line.

14.7 REDUCE OVERPRODUCTION

Overproduction is waste produced by producing and manufacturing more than the customer requires.

14.8 REDUCE DEFECTS

Waste of material, labor, and overhead due to defects results in scrap and rework.

14.9 INSPECTION OF INCOMING MATERIALS

The following steps will enable the reduction of material rejects, and consequently final product failure.

1. Ensure proper flow of parts and paperwork.
2. Inspect parts based on double sampling plan or according to deals with the supplier.
3. Inspect first part and report findings.
4. Issue a nonconforming report in case of any discrepancies in material specifications.
5. Retain nonconforming materials in a manufacturing record book.
6. Provide feedback to supply engineering section.

14.10 EMPLOYEE TRAINING

A company is only as good as its employees. In order for any company to succeed, it is absolutely necessary for the employees to attend training workshops and upgrade courses in order to be familiarized with new industrial developments to adopt a new technology to be used by the company. An employer must recognize that training employees is a long-term investment that would save the company money in many ways. Training employees is very important in product improvement. Employee training can include training workshops and training videos.

Employee qualification is when employees have met all identified skill and knowledge requirements for that position and have demonstrated their ability to produce a quality part with confidence in a safe and efficient manner.

Employee training is a great way to reduce production costs within a company. When employees are trained and updated with knowledge of new products and in using new jigs and devices, there is less chance of an error, which would lead to loss of time in the handling and manufacturing process. It also prevents severe injuries, rejects, lost materials, and reworks.

14.11 INSPECTION OF INVENTORY

An account of inventory must be kept and updated regularly in order to prevent a shortage of raw materials. It is essential to inspect inventory upon reception, in order to prevent manufacturer defects, which would cause defects in the finished products. This would reduce the cost of manufacturing, raw materials, and labor. It

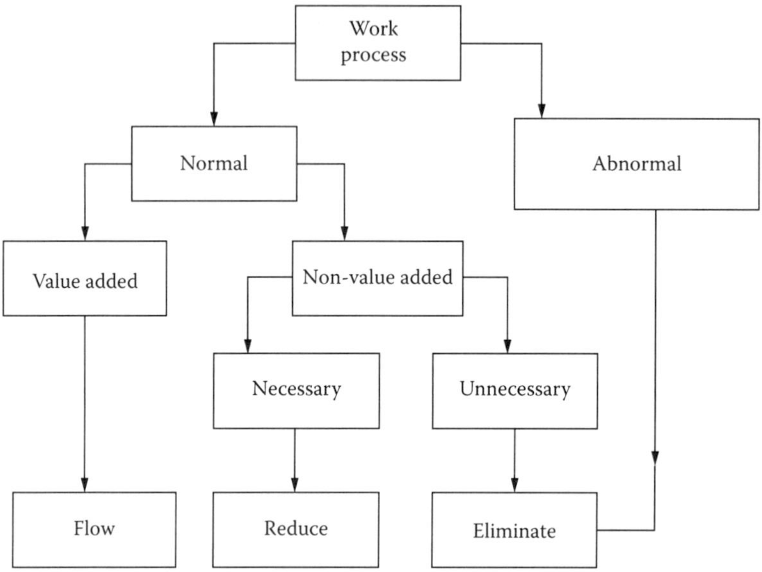

FIGURE 14.4 Identifying value and eliminating waste.

also reduces the cost overhead of energy consumption, maintenance, management, and insurance.

14.12 REDUCE REJECTS

In order to reduce rejects, it is important to identify value and eliminate waste. Proper analysis needs to be made in order to make any changes or adjustments. If research or studies are not conducted prior to making changes, there would be more harm than good achieved. In order to achieve zero defects, one most investigate both the process and how people interface with the process. Figure 14.4 shows how the processes are planned in a production line. Any unnecessary and abnormal processes must be eliminated.

14.13 USING ALTERNATIVE MATERIALS

Using alternative materials is another key method of product improvement. The research department of every company should evaluate new materials that may reduce cost while maintaining quality and durability of the product. Market research should be conducted in order to achieve consumer satisfaction of the new material before it is implemented into the production line and manufacturing process.

14.14 USING NEW JIGS, DEVICES, AND TOOLS

Changing the production line elements and handling facilities, such as trays, belts, and shelves, helps in overall product improvement. Using new jigs, devices, and tools will provide better job satisfaction. When you have good tools you do a better job.

14.15 USING AUTOMATION

The use of automated jigs, tools, and mechanical systems such as robots in the production line helps to increase the speed and effectiveness of any production line. Automated production lines may be more expensive and harder to maintain than manual labor, but the benefit is that they are very accurate and produce more standardized products. Automated production lines are fast and effective, and since there is less manpower used, there is less chance of injury. Automation cannot be used in some processes in the manufacturing of fiber optical devices and systems, which shows that in some cases, skilled hands can achieve better results.

15 Time Management in a Professional Environment

15.1 INTRODUCTION

A well-known quote ascribed to Benjamin Franklin, "time is money" is very true in today's business-driven environment. Even the most ordinary, everyday tasks require someone's valuable time. Therefore, it's not a secret why time is so valued by each and every individual in modern-day society who are striving to gain success.

The process of time management uses several tools and techniques to allow the successful planning of important tasks necessary to achieve the desired goals on time and on budget of an approaching deadline.

It is important to practice time management in one's personal as well as professional life to achieve effectiveness and efficiency while performing day-to-day tasks.

In order to keep up with the pace of the rest of the world, individuals and corporations must exercise time management to stay competitive and be able to meet customers' expectations.

15.2 TIME MANAGEMENT BEST PRACTICES

Effective time management requires learning and practicing certain tools and techniques in both one's personal and professional life. The following is a recommended list of multiple ways to assist in this endeavor.

15.2.1 Adopt a Positive Attitude

Pressures from both career and private life often distract us from following recommended practices for effective time management. Consider following the below subsections in order to adopt a positive attitude in the workplace.

15.2.2 Blame

Blaming life and work pressures for lack of time management will not help in organizing your day in order to achieve daily goals and work toward your life objectives. Setting daily lists of things to be done and following up on them is a key consideration in this aspect. Current computer and mobile technologies facilitate the creation of lists of tasks, and reminders assist you to outlay the day's tasks and record achievements in an easy and suitable manner.

15.2.3 Self-Assessment

Procrastination is a key factor in wasting time. Making a list of tasks and prioritizing them is the best way to manage the things that are required to be accomplished daily.

Internet, social media, and e-mail are many examples of distractions that can divert focus from what is necessary to be achieved. There is no problem to have some personal time as long as this does not supersede attending to higher-priority tasks.

Other distractions that can easily waste time at work include long coffee breaks and socializing and gossiping at the office water cooler. You can only manage time effectively by monitoring and limiting such activities, or compensating for the lost time by working later in the day after some colleagues have left. Also, start your workday at an earlier time when fewer colleagues are in the office.

15.2.4 Responsibility

It is said that the best-laid plans can change once the work starts. Managing change is a very important skill for time management. Having a prioritized list helps in managing changes by resetting the order of these tasks. Without it, you will be easily distracted waiting for the next task on the list.

15.2.5 Developing Specific Skills

Successful people achieve their goals with the help of time management skills. Identifying and practicing these skills moves you toward your life objectives faster. Some of these skills are as follows.

15.2.6 Prioritization Skills

Setting priorities focuses the mind on tasks and helps in achieving them faster. It is a key component of effective time management. The highest-priority tasks should require the utmost attention to complete them on time.

15.2.7 Planning Skills

Planning is key to success in life. The planning process involves identifying what has to be done (your objectives), how you are going to achieve them (your goals), and when everything has to be done (the tasks). Accomplishing tasks on time helps in achieving your goals and ultimately your objectives.

15.2.8 Delegation Skills

The ability to delegate tasks to others is a very valuable skill in general and in your work in particular. When you delegate responsibility to someone to do a certain task, you must ensure that you choose the right person for the job and give clear instructions of your expectations, such as what is to be accomplished and when you expect it. Otherwise, a lot of time can be wasted in redoing work. This can be costly, and divert you from focusing on other higher-priority tasks.

15.2.9 Organization Skills

Organization skills can be interpreted in multiple ways. It can mean the organization of the physical space, such as your office, desk, or home, but it can also mean organizing your relationships with your colleagues, coworkers, and family members. Identifying what has to be done and by whom is a skill that has to be mastered to be effective in life and work and improve your efficiency in achieving your goals and objectives.

15.3 ESTABLISHING DAILY PRIORITIES

Establishing daily priorities is a systematic process. The following processes and activities can be considered in the workplace in all daily work and activities to achieve smooth work hours, better results, and save time.

15.3.1 Identifying Goals from Objectives

Objectives are defined as high-level achievements that you are striving for in work and life. Goals are the steps that you need to take to achieve your objectives.

15.3.2 Identify Activities

Activities can be related to how you achieve your goals working toward accomplishing your objective. The first step in implementing a successful time management strategy is to identify activities and tasks that are to be carried out in order to perform a job.

15.3.3 Determine High- and Low-Return Activities

The second stage of the process requires identifying the high-return activities and focusing attention and effort into accomplishing them first. It is far more beneficial to spend effort completing these high-priority activities, rather than wasting time completing lower-priority activities that can be performed either toward the end or in between.

15.3.4 Consider Objectives

Identifying the objectives of a job helps put things in perspective. If a person knows what he or she is aiming toward, it helps him or her to focus all his or her energy and determination toward the achievement of the particular task.

15.3.5 Determine Urgency vs. Importance

Urgency is what defines important tasks that require attention in a workplace, as opposed to what takes place in personal lives. A distinction must be drawn between the two and focus should be laid upon urgencies in a work environment.

15.3.6 Rank Priorities (1-2-3 System)

Prioritizing tasks and ranking them allows things to be done in a step-by-step fashion based on their importance. This allows everyone in the organization to know the importance of the task and gets things done more sequentially.

15.3.7 Decide What to Do (ABC System)

Decide what to do using A, B, C, and so on. This system is similar to the ranking of priorities. This system works with daily, weekly, monthly, quarterly, and yearly activities. The tasks to be achieved are listed from most to least important. Each activity is named A, B, C, D, and so on. This makes it easy to follow up on the important or high-priority activities.

15.4 THE MYTHS

The following myths have considerable effects on success in the workplace.

15.4.1 Hard Work Leads to Success

While it may be true in some cases, there are no guarantees. Hard work and success are independent of each other. Achieving success by identifying your objectives and goals and then working hard toward achieving them can be a much better guarantee of success.

15.4.2 Keep Working and You Will Get More Done

That is a true fact of life, but what is important is that you focus your hard work on the right activities in the right order to achieve your goals and objectives. Activities and results have little in common. Often people are hardworking, but without having the right targets, it will not lead to successful results.

15.4.3 Efficiency Leads to Effectiveness

Efficiency is measured by how much of the right goals are achieved from the desired set. Effectiveness, in the meantime, is concerned with achieving the right tasks in the set order and timeline. Getting the proper training will result in working at a higher efficiency and good performance. A reduction in the cost of production will be achieved and also result in time saving, reject reduction, and better product.

15.4.4 Dedication and Commitment Are Essential to Success

Dedication and commitment to plans, employee satisfaction, and better customer service are essential to successful industrial and business activities. Customer satisfaction results in repeated sales, satisfied employees produce a better product, and having a clear business plan that is well communicated to the employees helps focus the organization on doing what is most important in achieving the common goals and objectives.

15.4.5 You Are the Best Person to Do the Job

An important part of delegating work is making sure that it is assigned to the most suitable person, who is most experienced to do that job. Training is a key goal to making sure that the employees are experienced in performing their daily tasks.

15.4.6 He or She Who Runs Away Lives to Fight Another Day

Taking responsibility for one's actions is a key success factor in business and personal life. Running away from problems instead of facing them head on builds a defeatist mentality and never solves the problem. Review the issues, analyze them, and break them down to smaller chunks if they are too big. Put a plan in action to remedy the issue. The most important thing is the understanding of what went wrong and to learn from the mistake. This is all part of quality management and key criteria in progressive and aggressive organizations.

15.4.7 Work Is Its Own Reward

No doubt some people are self-actualized in and through their work. Others are happy just to have a job. But few people still see work as an enabling experience.

Working hard and achieving great results becomes the reward by itself. Achievement should be rewarded and recognized at every level in the management system of any organization.

15.4.8 Work Was Never Intended to Be Fun

This is a false dichotomy—work should be fun. Successful people achieve this success because they enjoy what they are doing and therefore do not feel how hard they worked. Maybe sometimes one does not realize that he is enjoying what he is doing because of perfectionism. This way, he does not realize his achievements because he is always striving for better results. But that should not reduce the fact that he is enjoying his work.

For many, they are not blessed by this chance. They had to work hard for personal, family, and financial situations, but never found the job that gives them complete satisfaction. They believe that "work was never intended to be fun," but they keep doing it out of responsibility. For those, just having a job is a blessing. Global competition is putting more pressure on working-class people to hang on to their jobs and improve their performance. For these cases, finding a niche aspect of the job that is more in line with their personality and setting interim goals give them that missing sense of satisfaction.

15.4.9 It's My Way or the Highway

To think that there is only one best way of doing things blocks creativity and innovation and eventually leads to demoralization. Requesting feedback from employees and customers is one of the key principles of quality management science. Continuous improvement in production and service delivery has a major impact on growing and successful companies.

15.4.10 People Work Best under Pressure

It is a true fact, but only for a limited time. Working under pressure results in fatigue and can be a major factor leading to injuries and rework. This has been a way out for many of us, using it as an excuse for lack of planning. Running by the seat of our pants is a hit-or-miss scenario that can lead to disaster. Planning and prioritizing are the best antidotes to this disease.

15.5 TIME WASTERS

The following are time wasters. If they are avoided, time will be managed efficiently.

1. Overextended coffee and lunch breaks.
2. Extended and unproductive telephone conversations and texting.
3. Extended searches on the Internet.
4. Doing a task yourself that could have been delegated to someone with higher qualifications for the job.

5. Spending too much time on personal and other nonproductive conversations.
6. Letting others use up more of your time than they are entitled to. The reverse of this is for you to be inconsiderate of other people's time by overstaying an appointment to discuss personal or inconsequential matters.
7. Lost time due to waiting for others because of your poor scheduling or your failure to take more with you to do while waiting.
8. Permitting late-afternoon drag to slow you down.
9. Spending too much time on a problem or project when asking for advice would save time and result in a better solution.

15.5.1 Interruptions

Interruptions of all sorts, such as telephone conversations, human resources, subordinates, superiors, colleagues, and friends, are probably one of the greatest obstacles to effective time usage. Someone effective in his or her position must concentrate all energy, creativity, and analytical or decision making on a given problem or project. If someone is interrupted, the chances are that will achieve half the results that would be achieved otherwise.

The following steps could help to reduce or eliminate interruptions:

1. Establish a quiet working hour policy.
2. Establish a closed-door policy.
3. Use a meeting room anywhere to accommodate the participants.
4. Tell colleagues that you are busy working and times are scheduled on a board or on the door office.
5. Schedule working hours, meetings, and consultations.

15.5.2 Correspondence

Handling mail and Internet memos can take up a considerable amount of time. Here are a few techniques that will help to organize correspondence time:

1. Have human resources screen the incoming mail.
2. Use the ABCD system. The mail should immediately be divided into four groups:
   A group—top priority, act quickly
   B group—delegate it
   C group—place in a pending file and act when time permits
   D group—throw it out
3. Deal with pending files at a specific time each day.
4. Choose the most appropriate time to deal with correspondence.
5. Use the phone instead of writing a letter. Use the phone when there is no decision, commitment, or financial agreement.
6. Answer letters only when necessary. Not all letters require an answer.
7. Use all communication systems in the workplace and do not use your personal system.

8. Organize an effective filling system.
9. File documents, memos, and letters on a daily basis.
10. Send any catalogs, brochures, and leaflets to the related destiny.
11. Review the filing system at least once a year.

15.5.3 Meetings

Many people spend about 25 to 30% of their time in a meeting. Many believe that these hours are not always as productive as they might think. To improve the meeting time, the following is recommended:

1. Reduce the number of participants in the meeting by including those who are important.
2. Calculate the cost of a meeting.
3. Hold the meeting when necessary.
4. Choose the time carefully. The best times for a 1-hour meeting are:
   9:00 meeting that will end with a coffee break
   11:00 meeting that will end with the lunch break
   16:00 that meeting will end with the workday
5. Choose the right kind of meeting room.
6. Prepare an agenda for the meeting.
7. Set time limits per topic on the agenda.
8. Manage the time effectively.
9. Follow up the meeting systematically.

15.5.4 Meeting Rules

The following is the list of meeting rules:

1. Everyone is on time for the meeting.
2. Set an agenda.
3. Set time limits for each item on the agenda.
4. Have a timer to keep the meeting on schedule.
5. Put any other items on the "deferred to later" list.
6. Only one person talks at a time.
7. Do not interrupt anyone that is talking.
8. Do not criticize anyone's ideas in the meeting. Make a separate meeting to discuss new ideas.
9. Set agenda topics for the next meeting.
10. Finish the meeting on time.

15.5.5 Communications

Communications, such as phone, e-mail, fax, and online, allow us to save a great deal of time by cutting down on trips, visiting, meetings, and correspondence. But also, they can become addictive. The following is important to be considered:

1. Speak to someone on an important issue.
2. Use voice mail prompts to get the necessary information.
3. Make all calls at a specific time during the day. The best times are:
   8:30 to 9:00
   11:30 to 12:00
   13:30 to 14:00
   16:30 to 17:00
4. Make calls short, to the point, and confirm the outcomes at the end.
5. Never tolerate phone calls and e-mails in a meeting.
6. Use e-mail or fax when sending documentation when five people or more have to be reached.
7. Plan calls and e-mails before making them.
8. Hold a conference meeting when traveling time, cost, and more than two people are involved.

15.5.6 Top 10 Time Management Tips

1. Write down your goals. Use the smart formula: make them specific, measurable, appropriate, realistic, and time-bound.
2. Create a mission statement for your life. Like a large company, your mission should state what you stand for and where you want to be.
3. Every day, divide your tasks into A, B, C, and D priorities. Always start with an A, even if you can only accomplish a small part of it.
4. Block off time for activities that are important. Make an appointment with yourself, and don't let anyone schedule anything for the time you have put aside.
5. Have the courage to say no. Don't try to please others all the time.
6. Always start meetings on time.
7. Ask people who interrupt you if you can meet them later. Set a time when you can visit them in their office, so you can leave when you are finished.
8. Avoid procrastination by completing unpleasant tasks first. You might find they turn out to be not so bad after all.
9. When leaving phone messages, tell others exactly why you are calling. This will help them when they call you back and ultimately save time for both of you.
10. Create time for balance in your life. Set aside time for family, fitness, social, educational, and spiritual needs.

15.6 TIME MANAGEMENT STRATEGIES

Time management strategies are based upon the creation of action plans to implement the task lists. These tasks have to lead to accomplishing goals, leading to achieving personal and organizational objectives. Therefore, these objectives have to be written down, well communicated, and understood by everyone involved.

Follow techniques to identify, to prioritize, and to estimate the duration and that the way the plan will be implemented is the best approach to achieve success in all endeavors.

This approach increases the efficiency of doing the work and improves the effectiveness of effort in achieving goals and objectives. Implementing the plan and managing the change is the best way not to fall under the effect of all the myths surrounding the definition of success on personal and corporate levels.

16 Internal, External, and ISO Auditing

16.1 INTRODUCTION

Auditing is an independent practice, objective assurance, and consulting activity designed to add value and improve the operation of a production line. It is conceptually similar in many ways to financial auditing by public accounting firms, quality assurance, and banking compliance activities. Auditing also helps a company to accomplish its objectives by bringing an organized, disciplined approach to evaluate and improve the effectiveness of management, control, and safety processes in a production line.

There are two ways to perform auditing: internal and external. Internal auditing is completed by a group within the company. External auditing is executed by an independent, certified company. There are many advantages when performing internal and external auditing. The most important one is to keep a production line in perfect working condition and comply with all local and international standards. This leads to consequent cost savings and other benefits in the long run. In this chapter, general information about auditing practices and performance will be presented, especially when auditing a production line.

16.2 INTERNAL AUDITING

Internal auditing helps improve a production line's performance by providing insight and recommendations based on data collection, analyses, and assessments of data processes. Internal auditing provides valuable information to management and engineers as an objective source of independent assessment and advice. Professional engineers from the company perform internal auditing activities in their production line or any line in the company. Internal auditing a production line within a company is comprehensive and may involve topics such as manufacturing procedures, quality of products, quality assurance, efficiency, and employee compliance with laws, regulations, and standards. The auditors direct management and engineers regarding assessments in order to improve the execution of their responsibilities. The financial and management audit is not the responsibility of the professional engineers in internal auditing activities. Internal auditors are not responsible for the execution of company activities. Internal and external auditing have ready-made templates to be used in auditing production processes. These templates are updated, depending on the laws, regulations, and standards in the field. There are general and specific templates, depending on the type of production line and product.

16.3 EXTERNAL AUDITING

Normally a special, professional company is tasked to perform the external auditing activities for each field. Usually, external auditing is performed after internal auditing. External auditors do not take into consideration internal auditors' findings in their report; instead, management takes care of the internal auditors' report. Managers, engineers, and supervisors take proper action before external auditors start their activities auditing a production line. External auditors start by auditing a production line's performance and provide comprehensive recommendations based on analysis and assessment of the data. Similarly, internal auditors provide valuable information to management and engineers as an objective source of independent assessment and advice. External auditors auditing a production line are comprehensive, and may involve many topics, such as manufacturing procedures, quality of products, quality assurance, efficiency, and employee compliance with laws, regulations, and standards. At the end of auditing activities, an auditing report is produced for management and the engineers, showing the assessment's findings. External auditors are not responsible for the corrective actions listed in the report's recommendations. There are general and specific forms, depending on the production line and product.

16.4 AUDITING ACTIVITIES

Auditing activities are primarily directed at evaluating a production line's components, documentation, and performance. Auditor activities are generally defined as a predefined process, designed to provide reasonable assurance regarding the achievement of the following core objectives for which all companies strive:

- Effectiveness and efficiency of operations
- Consistency of documentation and applications
- Device and system certifications
- Employees understand manufacturing operations
- Compliance with industrial laws and regulations

16.5 MANAGEMENT'S AND ENGINEERS' ROLES

Managers and engineers are responsible for manufacturing and control, which are comprised of the following critical manufacturing components:

- Control of manufacturing environment
- Risk assessment
- Focus control activities
- Information and communication
- Monitoring activities
- Continuous update actions

Managers and engineers establish policies, procedures, and practices to help the organization achieve the specific objectives listed above. Auditors perform audits to evaluate whether the components of management in manufacturing are present and operating effectively. They will provide recommendations for improvement in case such components are missing.

16.6 AUDITING ANNUAL PLANNING

Companies require auditing certificates for many reasons. The most important reasons are to comply with industry requirements, insurance policies, and customer satisfaction, and also to stay competitive in the market. Specifically, managers and the board of directors are focused on auditing efforts. This focus or prioritization is part of the annual or multiyear audit planning strategy of the company. The audit plan is typically planned by management for the review and approval of the audit committee or board of directors. Any internal auditing activity is generally conducted as one or more discreet assignments.

16.7 EXECUTION OF AUDITING ACTIVITIES

A typical auditing assignment involves the following steps:

1. Establish and communicate the scope and objectives for the audit to suitable management.
2. Study the manufacturing processes under review.
3. Auditing action includes objectives, measurements, reviews, and employee interviews.
4. Describe the difficulties facing the company activities within the scope of the audit.
5. Identify management practices in auditing the components of control used to ensure that each element is properly controlled and monitored. An audit checklist can be created and is a helpful tool to identify common desired controls in the specific production line being audited.
6. Develop and execute a sampling and testing approach to determine whether the most important elements and controls are operating as intended.

7. Create an audit report to identify findings and a list of problems.
8. Create a communication channel to negotiate action plans with cooperation from management and engineers to address the problems.
9. Create a follow-up plan on reported findings with designated individuals. The audit group maintains a follow-up database for this purpose.

16.8 AUDIT REPORTS

Auditors typically issue reports at the end of each audit activity that summarize their findings. The report also includes recommendations and any responses or action plans from management.

The report may have an executive summary, a body that includes the specific issues or findings identified and related recommendations or action plans that must be implemented, and appendix information, such as detailed graphs and charts or process information. An automated online report can be generated for easy access on many electronic devices.

Each audit report may contain five elements, sometimes called the 5C's:

Condition: To identify a particular problem or problems.

Criteria: Determining whether local and international standards, regulations, and company policy are met and to also identify deviations.

Cause: The reasons why some problems occur and what the reasons are.

Consequence: Following up on the report recommendations by identifying a professional solution and implementing it urgently in order to avoid negative impact on the company's reputation.

Corrective action: Some recommendations require urgent corrective actions. Others can wait sometimes. The company should establish a preventative action plan for these activities.

The recommendations in the audit report are designed to help a company achieve effective, risk-free, and efficient performance in controlled processes in line with operational objectives and legal compliance objectives.

16.9 QUALITY OF THE AUDIT REPORT

The audit report should be comprehensive and also adequate for each auditing activity. The following are quality elements to be found in an audit report:

Objectivity: The assessments, findings, comments, and recommendations expressed in the report should be objective and neutral.

Clarity: The language used in the report should be simple, short sentences, and straightforward.

Accuracy: The information contained in the report should be accurate and confirmed from professionals in the field.

Brevity: The report should be short, presenting the findings and recommendations straightforwardly.

Timeliness: The report should be released on time, immediately after the audit is completed. The date for the release of the audit report should be as outlined by the regulations.

16.10 AUDIT REPORT STRATEGIC PLANS

Auditors may also develop functional strategies as described in multiyear strategic plans. Professional guidance on building a strategic audit plan was put in place for future application. A key aspect of developing strategic plans is to understand the expectations of stakeholders, employees, and industrial demands. Strategic plans guide the company to be ready in order to better face risks and problems. Specific topics considered in strategic planning include the following:

Scope and emphasis: Strategic plans may be involved in addressing risks related to financial, industrial development, operations, legal and regulatory compliance, and company strategy.

Portfolio of services: Strategic plans may provide traditional audit assurance across the risk spectrum, as well as project support in a variety of areas, such as project management, data analysis, and monitoring initiatives.

Competency development: Strategic plans may provide management around the scope and service portfolio to determine need for improvement, expectation, the hiring of specific skills, and the creation of training programs.

Technology: Strategic plans highlight some shades on the evolution of a new technology. Implementing new technology and using automation in manufacturing keep the company active in the market and will therefore reduce cost.

16.11 THE DIFFERENCE BETWEEN INTERNAL AND EXTERNAL AUDITS

While sharing some characteristics, internal and external auditing have very different objectives. These are explained and presented below:

1. Regardless of the type of audit, the starting point is always the objectives and goals of the company.
2. Identify similar risks to achieving audit objectives of the company.
3. The responsibility of the internal auditor is different from that of an external auditor.
4. The types of work performed by external versus internal auditors differ. The differences are primarily in the type of objective and type of risk assessed by the auditors.

5. The auditors in both audits plan the audit effort around the area assigned for the audit to achieve the objectives.
6. There are companies that have various objectives for internal and external audits.
7. Once an objective is identified, the auditor can start working on auditing the production line.
8. Internal auditors who are members of a company are subject to the same code of ethics and professional code of conduct as external auditors. However, they differ primarily in the relationship to the entity they audit.
9. Achieving improvements is fundamental to the purpose of internal auditing. It is done by advising, coaching, and facilitating in order to not undermine the responsibility of management. An external auditor, however, has a duty to report findings and problems.
10. The overall audit thought processes for both types of auditors are similar.

16.12 THE ROLE OF THE INTERNAL AUDITOR

The following are internal auditing roles:

- Internal auditors are typically internal employees of the company.
- Internal auditors, though generally independent of the activities they audit, are part of the company they audit and report to management.
- Internal auditors are responsible for manufacturing procedures, devices and systems certificates, operations, and quality control of a company.
- The internal audit activity must assist the company in maintaining effective controls by evaluating their effectiveness and efficiency regarding the following:
  - Reliability and integrity of manufacturing processes and controls
  - Effectiveness and efficiency of operations
  - Application of standards
  - Compliance with laws and regulations
- Internal auditors should ascertain the extent to which operating and program goals and objectives have been established and conform to those of the company.
- Internal auditors should review manufacturing operations and controls to determine the extent to which results are consistent with established goals and objectives to see whether operations and controls are being implemented or performed as planned.
- The internal auditing objectives and goals vary from one company to other. The main objectives are in the following four categories: strategic, operations, reporting, and compliance.

16.13 THE ROLE OF THE EXTERNAL AUDITOR

The following are the external auditor roles:

- External auditors are external to the company and are not employees of the company.
- External auditors may be liable to third parties who are damaged by making decisions based on information in audited reports.
- The external auditor is primarily concerned with the objectives of auditing and reporting to the management.

16.14 INTERNATIONAL ORGANIZATION FOR STANDARDIZATION (ISO)

Product quality is a topic of interest throughout the global market. As the market environment evolves because of advancements in telecommunication and transportation technologies, the need for a rapid response to global developments becomes more critical. In the global market, product characteristics (expressed in quality terms) will be a major basis for such market negotiations. Manufacturers, retailers, distributors, venders, and consumers will all need to have a common understanding of what constitutes acceptable product quality.

Striving for better quality worldwide has led to the need for unified international quality standards. The International Organization for Standardization (ISO), located in Geneva, Switzerland, has developed general quality guidelines known as ISO 9000. The ISO is a special international agency for standardization of the national organization from several countries. The ISO family has over 19,500 standards touching almost all aspects of daily life. The ISO 9000 family of quality management systems standards is designed to help companies ensure that they meet the needs of customers and other stakeholders while meeting statutory and regulatory requirements related to a product. ISO 9000 deals with the fundamentals of quality management systems, including the eight management principles upon which the family of standards is based. ISO 9000 deals with the requirements that companies wishing to meet the standard must fulfill. ISO 9000:2005 (the actual standard name) explains the fundamentals of quality management systems (like ISO 9001), and the vocabulary used in the ISO 9000 standards ISO certification provides independent confirmation that companies meet the requirements of ISO 9000. Over a million companies worldwide are independently certified, making ISO 9000 one of the most widely used management tools in the world today. Despite widespread use, the ISO certification process has been criticized as being wasteful and not useful for all companies.

ISO 9000:2005 focuses on what makes up a quality management system (QMS). In a sense, it provides the framework for a group of standards that govern how a consistent, training-driven quality system should function. This is whether it is serving a wide variety of industries (like ISO 9001) or a more specific industry, like the automotive standard ISO/TS 16949. In writing a standard to govern how a quality system should broadly function, ISO lays out the basics of such a system and provides definitions and concepts that can be applied to the ISO 9000 grouping of quality systems.

The fundamentals of quality management systems used in the ISO 9000 standards are:

- Describe the fundamentals of the QMS in the ISO 9000 family.
- Define the related terms in the ISO 9000 QMS.
- As used as guidance when implementing ISO 9000 standards.
- Does not have any requirements, so an organization does not certify to ISO 9001.
- Is scheduled to be updated in the future.

This standard is useful to a fairly specific set of users. It may include companies that may be seeking to apply multiple standards, end users that want to understand how the quality of their products or services are maintained, those who wish to gain a broader understand of terms and definitions used in ISO-constructed quality systems, those who must administer an ISO quality management system in some way, and those seeking to develop standards of their own that might need to harmonize with ISO 9000 from the overall family and grouping perspective.

16.15 BENEFITS OF ISO 9000 STANDARDS

ISO standards bring technological, economic, and societal benefits. They help to harmonize technical specifications of products and services, making industry more efficient and breaking down barriers to international trade. Conformity to ISO 9000 helps reassure consumers that products are safe, efficient, and good for the environment.

ISO standards have developed materials describing the economic and social benefits of standards, the ISO materials. They are intended to be shared with decision makers and stakeholders as concrete examples of the value of standards.

The repository of studies on economic and social benefits of standards provides an insight of the approaches and results of the studies undertaken by different authors, such as national and international standards bodies, research institutes, universities, colleges, and other international agencies.

ISO standards are strategic tools and guidelines to help companies tackle some of the most demanding challenges of modern business. They ensure that business operations are as efficient as possible, increase productivity, and help companies' access new markets.

ISO standards give better-quality management and help to meet customer needs. ISO standards also help improve quality, enhance customer satisfaction, and increase sales.

ISO standards help optimize operations, and therefore improve the bottom line and cost savings. They improve operational performance by cutting errors and increasing profits.

ISO motivates and engages staff with more efficient internal processes.

ISO standards help prevent trade barriers and open up global markets. They also broaden business opportunities by demonstrating compliance. ISO standards help increase productivity and competitive advantage around the world.

ISO standards help reduce negative impacts on the environment and health.

ISO benefits business and governments. ISO standards draw on national and international expertise and experience, and are therefore a vital resource for governments when developing regulations.

16.16 BENCHMARKING FOR ISO 9000

The objective of a company in international benchmarking is to match the best in the global market. Benchmarking is a process where target quality standards are established based on the best examples in the market. Benchmarking implies learning from the best. The premise of the benchmarking is that if an organization follows the best-quality examples, it will become one of the best in the market. A major objective of benchmarking is to identify deficiencies in quality and attempt to remove them. The essential elements of benchmarking are as follows:

- Monitor prevailing standards in the market.
- Identify who is meeting the national and international standards.
- Exchange updated information with market leaders.
- Measure and determine current market performance level and future expectation.
- Analyze bottlenecks and gaps in performance.
- Continuously strive to close the bottlenecks and gaps.
- Make a future plan.

16.17 RESPONSIBILITIES OF MANAGEMENT

Management responsibilities for the ISO 9000 registration process can be summarized into the following important elements:

- Designating a management representative to lead the cause of ISO 9000. There are roles to be followed by the management representative.
- Establishing a steering committee to communicate with management. There are roles to guide the steering committee.
- Being familiar with the general aspects of ISO 9000 standards and be sure that someone in the organization is up to date with the details.
- Knowing, as an ISO 9000 representative, what sections apply to specific functional areas.
- Ensuring that work groups comply with needs and changes required by ISO 9000 as determined by the steering committee.
- Setting the ultimate goal of expanding ISO 9000 standards compliance throughout the organization.
- Providing the time and resources required for ISO 9000 standards preparation, audit, and follow-up.

16.18 ISO 9000 EMPLOYEE EXPECTATIONS

Employee expectations can be summarized as follows:

- Be aware of ISO 9000 and how it affects your performance and working area.
- Follow compliance plan and procedures in your working area.
- Correct inaccurate documents in your working area.
- Follow updated documented work instructions, such as manufacturing, testing, shipping, quality control processes, and safety instructions.
- Cooperate with the steering committee, management representative, quality control, and auditors.

16.19 TIPS FOR A SUCCESSFUL ISO 9000 PROCESS

Even though the ISO standards appear to give rigid guidelines, the implantation is really flexible. Tips for a successful ISO 9000 process are as follows:

- Keep the approach simple and flexible when applying the regulations and roles.
- Develop a brief work plan and detailed procedures.
- Make necessary adjustments to processes and instructions in the workplace.
- Use a benchmarking approach to measure other compliances with ISO 9000.
- A quality employee should chair ISO 9000 activities.
- Use outside experts and resources to support ISO 9000 activities.
- Involve everyone in the organization in ISO 9000 activities.
- Be flexible for applying ISO 9000 as a rigid set of standards.
- Stay close to ISO 9000 requirements.
- Keep an eye on market developments.
- Learn, train, practice, and keep up.

16.20 IMPACT OF TECHNOLOGY CHANGES AND ISO 9000

Fast technology developments and changes can adversely affect compliance with ISO 9000 standards. ISO 9000 must adapt to new technologies and change and update operating procedures and documentation accordingly. Photonics, microcomputers, the Internet, and medicine are very good examples. Technology changes frequently necessitate new operating procedures. Work process documentation, quality control, and safety procedures must be frequently updated in such an industry to keep up with the technological changes. Many fields are in highly competitive industries in which most companies are actively pursuing ISO 9000 certification.

17 High-Tech Manufacturing Safety

17.1 INTRODUCTION

Our lives are filled with hazards created by electrical power supply, lasers, chemicals, and a diversity of equipment used in manufacturing lines. While manufacturing engineers will learn to identify life's hazards and how to protect themselves and employees, employees will learn to take care of their health and safety while working in the manufacturing environment. A safer and healthier learning and working environment should be created so that employees have the opportunity to live safer and healthier.

The following list of safety reminders is a brief compilation of generally accepted practices and should be adopted or modified to suit the unique aspects of each working environment and local and provincial or federal codes. The intent of this chapter is to stimulate thinking about important safety considerations for employees in manufacturing.

17.2 PERSONAL PROTECTIVE SAFETY EQUIPMENT (PPSE)

Personal protective safety equipment (PPSE) is essential in the protection of workers, as shown in Figure 17.1. PPSE can include goggles, clothing, helmets, or other equipment to protect a worker from hazardous materials or environments. In high-tech manufacturing and laser facilities in particular, a major piece of PPSE needed is laser safety goggles. Laser goggles can vary depending on lasers used. The type of glasses needed to be worn can depend on the wavelength, power, and type of laser. When using any laser over a Class 1 (under 0.4 mW), glasses are recommended; anything over Class 2 (over 1 mW), glasses are a necessity. Laser safety will be explained in detail in this chapter. It is important to wear the proper glasses for the correct laser, as some glasses that were designed for another laser will have no protection if used with another of a different wavelength and power.

Special clothing also might be needed while operating certain laser systems. Some lasers with ultraviolet rays may cause damage to the skin. They can cause increased signs of aging and a higher risk of getting skin cancer. For this reason, around any ultraviolet rays for long periods of time it is a good idea to wear protective clothing.

Another form of PPSE that is common in high-tech manufacturing and laser facilities is gloves. Gloves are essential to protect the fingers against chemicals, fiber optic cables, sensitive devices, systems, etc. For example, the core of the fiber cable is pure glass. This can be very harmful if a piece gets broken and enters the bloodstream, as it could end up in the heart or brain and cause serious injury or death. It is very important for gloves to be worn whenever handling any type of fiber optic cables, sensitive devices and systems, chemicals, biological and lab materials, etc.

FIGURE 17.1 Personal protective safety equipment (PPSE).

17.3 ELECTRICAL SAFETY

The importance of electrical safety cannot be overstressed. Electrical accidents can result in property damage, personal injury, and sometimes death. Ensuring electrical safety in manufacturing facilities is important for employees. Employees can learn to have a healthy respect for electricity, and to spot potential electrical hazards anywhere they may be. This does not mean that one should fear electricity: just use it properly and wear personal protective equipment.

17.3.1 Fuses/Circuit Breakers

The most common protection against property damage from circuit overloads (too much current) and overheating is the use of fuses and circuit breakers. All electrical circuits in manufacturing facilities are required to be protected by these means. When too much current flows in a circuit, the circuit becomes hot and could melt the wire insulation, emit caustic fumes, and start a fire. An overload may also burn out and damage devices and instruments. Electronic equipment commonly has fuses to protect the components from overloads. A fuse is essentially a short strip of metal with a low melting point. When the current in a fused circuit exceeds the fuse rating, for example, 3, 5, 15, or 20 A, the heat melts or vaporizes the fuse strip. The fuse blows and the circuit is opened. Figure 17.2 shows types of fuses used in most electric instruments in the manufacturing facilities.

Fuses and circuit breakers should be the correct current rating for the circuit. If the correct rating is unknown, a certified electrician can identify and label it. Always replace a fuse with another of the same rating. Determine the reason why a fuse blew or a circuit breaker tripped, before replacing or resetting. Figure 17.2 also shows a few types of fuses and a power supply. The fuse will need to be plugged in on the back of the power supply. Plug in the electrical cord, then turn on the key switch on

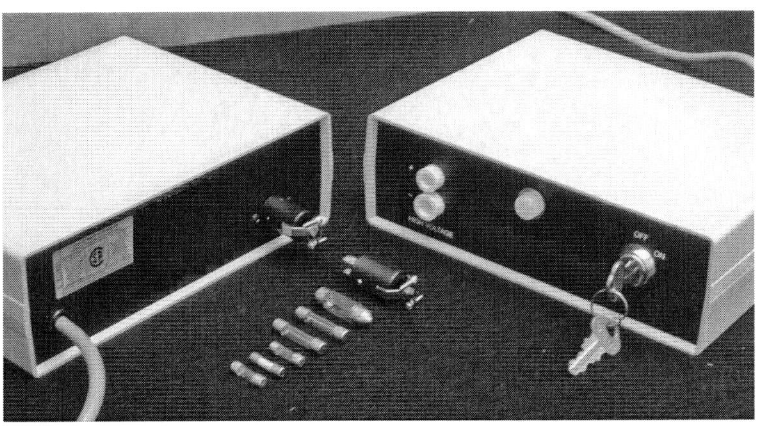

FIGURE 17.2 Types of fuses.

the front panel to turn the power supply on, as shown in Figure 17.2. After finishing with the power supply, remember to turn off the key and unplug the fuse.

A common problem is that the insulation on wires may become worn, for example, on an extension cord, device wire, or instrument cord. If bare wires touch each other, or if a high-voltage or hot wire touches ground, this is called a short circuit, since the path of the circuit is effectively shortened. A low-resistance path to ground is created, causing a large current, which blows the protecting fuse.

It is more common now to use circuit breakers instead of fuses in large equipment and houses, as shown in Figure 17.3. If the current in a circuit exceeds a certain value, the breaker is activated, and a magnetic relay (switch) breaks or opens the circuit. The circuit breaker switch can be reset or closed manually.

In either case, whether a circuit is opened when a fuse blows or when a circuit breaker trips, steps should be taken to remedy the cause. Remember, fuses and circuit breakers are safety devices. When fuses blow and open a circuit, they are indicating that the circuit is overloaded, shorted, or another electrical problem. In any case, a certified technician must investigate the source of the problem.

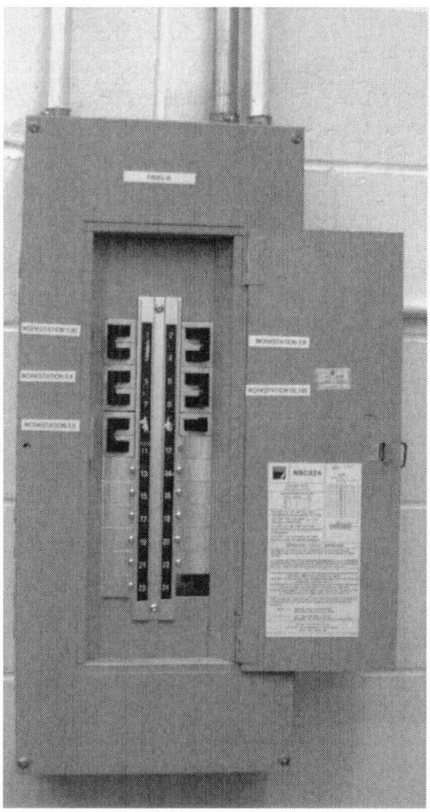

FIGURE 17.3 A circuit breaker panel.

FIGURE 17.4 Examples of switches.

17.3.2 ON/OFF Switches

Figure 17.4 shows samples of ON/OFF switches, which are used in computers, lighting systems, and instruments.

17.3.3 Plugs

Switches, fuses, and circuit breakers are always placed in the hot (high-voltage) side of the line, to interrupt power flow to the circuit element. However, fuses and circuit breakers may not always protect from electrical shock. To prevent shock, a grounding wire is used. The circuit is then completed (shorted) to ground, and the fuse in the circuit is blown. This is why many electrical tools and appliances have three-prong plugs, as shown in Figure 17.5. In the wall receptacle, this connection runs to ground.

When trying to plug in a two-prong plug and the plug does not fit, do not use any force to plug it in. Turn the plug over and try again. Figure 17.6 shows a two-prong plug. One of the prongs is bigger than the other. This is called a polarized plug. Polarizing in the electrical sense refers to a method of identification by which proper connections can be made. The original intent of this type of plug was as a safety

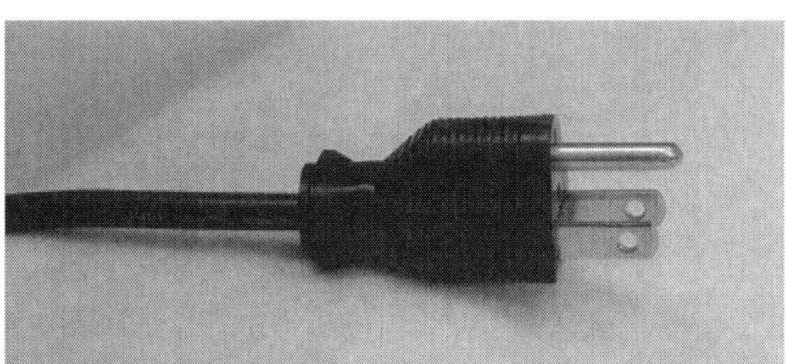

FIGURE 17.5 A three-prong plug.

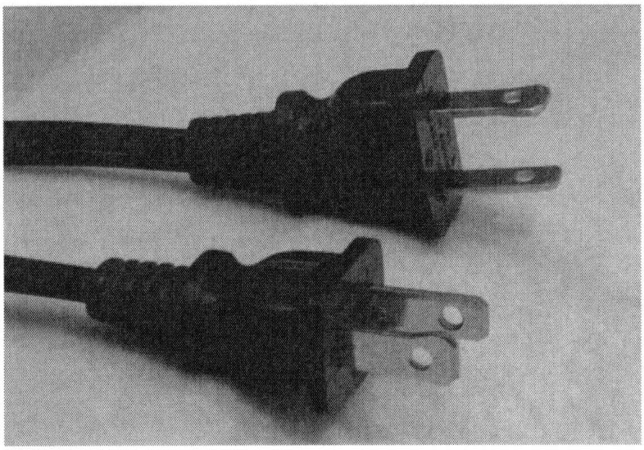

FIGURE 17.6 Two-prong plugs.

feature. The small slit in the receptacle is the hot side, and the large slit is the neutral or ground side, if properly connected. The housing of an appliance could then be connected to the ground side all the time via a three-prong plug. A receptacle or appliance that is not wired (polarized) properly can be dangerous. The polarization is ensured with a dedicated third grounding wire as in a three-prong plug system, which is the accepted safety system. The original two-prong polarized plug system remains as a general backup safety system, provided it is wired properly.

Ensure the plug type fits the receptacle type. Never remove the ground pin (the third prong) to make a three-prong fit into a two-conductor outlet; this could lead to an electrical shock. Never force a plug into an outlet if it doesn't fit. Plugs should fit securely into outlets. Avoid overloading electrical outlets with too many devices.

17.3.4 Wall Outlets

Figure 17.7 shows a wall outlet, which is used to connect computer and extension cords. Avoid using wall outlets that have loose-fitting plugs. They can overheat and lead to fire. Ask a certified technician to replace any missing or broken wall plates.

17.3.5 Cords

Ensure the cords are in good condition. Check cords for cut, broken, or cracked insulation. Protect flexible cords and cables from physical damage. Ensure they are not placed in traffic areas. Cords should never be nailed or stapled to the wall, table, baseboard, or another object. Do not place cords under a device, computer, or rest under any object. Cords can create tripping hazards and may be damaged if walked upon. Allow slack in flexible cords to prevent tension on electrical terminals.

Check that extension power bars are not overloaded, as demonstrated in Figure 17.8. Additionally, extension power bars should only be used on a temporary basis; they are not intended for use as permanent wiring. Ensure that the extension power bars have safety closures.

FIGURE 17.7 A wall outlet.

(a) Extension power bar is overloaded.

(b) Extension power bar is not overloaded.

FIGURE 17.8 An extension power bar.

17.3.6 Ground Fault Circuit Interrupters

Ground fault circuit interrupters (GFCIs) can help prevent electrocution. They should be used in any area where water and electricity may come into contact, especially near a sink or basin. Water and electricity don't mix; they create an electrical shock. When a GFCI senses current leakage in an electrical circuit, it assumes that a ground

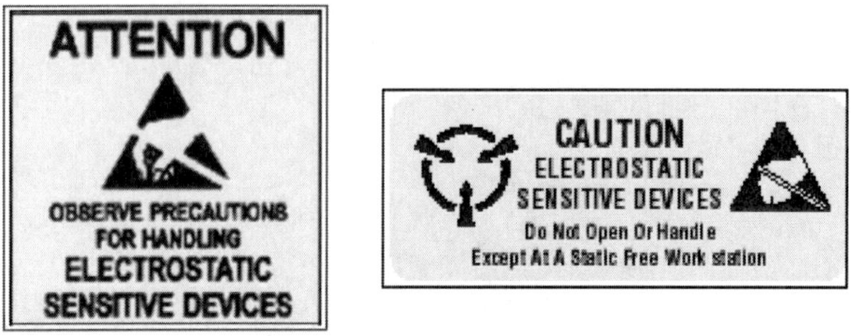

FIGURE 17.9 Electrostatic discharge warning symbols and signs.

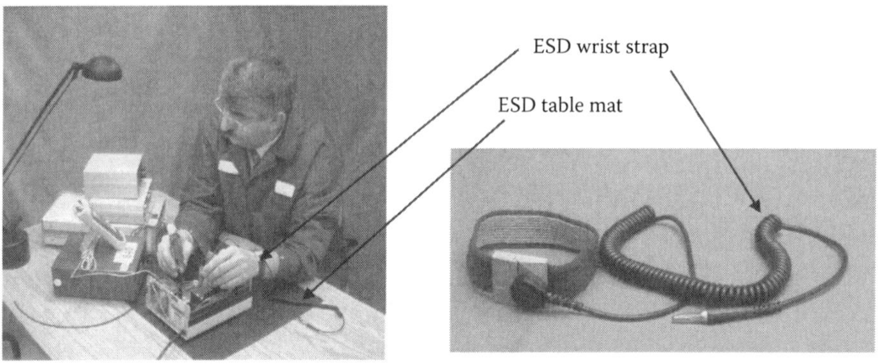

FIGURE 17.10 Working with ESD wrist strap and table mat.

fault has occurred. It then interrupts power fast enough to help prevent serious injury due to electrical shock. GFCIs should be regularly tested according to the manufacturer's instructions to ensure they are working properly. Some benches are connected to true ground to be electrostatic discharge compliant. This is very important for sensitive devices and equipment that is very sensitive to electrostatic discharge. Figure 17.9 shows electrostatic discharge (ESD) warning symbols and signs.

Figure 17.10 shows an ESD wrist strap and table mat used in handling an ESD-sensitive device. The straps and mats should be connected to the true ground before handling a sensitive device. The strength of the charge on a human body is enough to destroy an ESD-sensitive device. Each person should discharge his or her electrostatic charge before entering an environment sensitive to ESD. The discharge devices are usually located at the entrance of sensitive areas. An ESD heel strap is also available to wear when handling devices and walking in an environment sensitive to ESD.

Figure 17.11 shows ESD bags used to package devices sensitive to ESD. The bags are available in various sizes and have printed labels.

17.4 LIGHT SOURCES

The wattage rating of bulbs should be checked for all bulbs in light fixtures, table lamps, light sources, etc., to make sure they are the correct rating for the fixture.

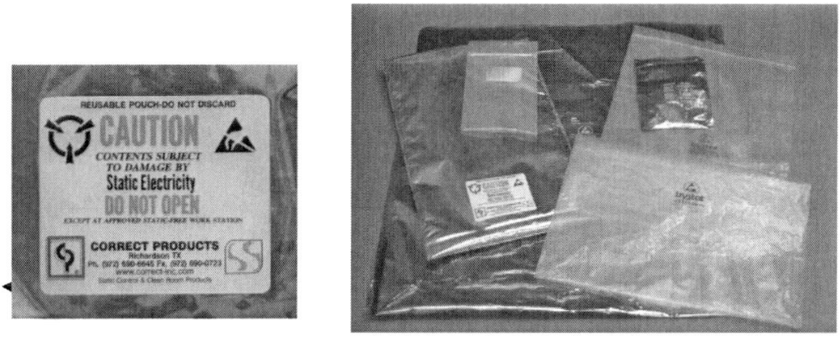

FIGURE 17.11 ESD bags.

Bulbs must be replaced with another of the same wattage rating; bulbs must have no higher wattage than recommended. If the correct wattage is unknown, check with the manufacturer of the fixture. Ensure the bulbs are screwed in securely; loose bulbs may overheat. Different gas light sources (hydrogen, mercury, neon, etc.), as shown in Figure 17.12, are used in manufacturing facilities for light loss measurements, spectrometers, and optical applications. These lamps operate at much higher temperatures than standard incandescent and florescent light bulbs. Never place a lamp where it could come in contact with any combustible materials or near the skin. Be sure to turn the lamp off before leaving the area for an extended period of time. Note that laser light sources have special provisions and precautions to operate them.

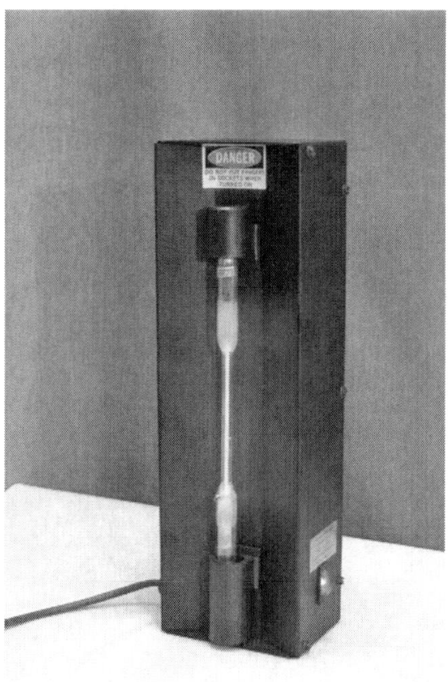

FIGURE 17.12 A mercury light source.

17.5 DEVICES AND EQUIPMENT

If a device or equipment repeatedly blows a fuse, trips a circuit breaker, or if it has given you a shock, report the incident immediately to your supervisor/instructor. Unplug the device and remove it to have it repaired or replaced.

17.6 AUDIOVISUAL AND COMPUTER PERIPHERALS

Audiovisual and computer equipment must be checked and kept in good working condition. Ask the technician to load the printer with paper and replace the toner. Report the faulty equipment to the technician to be repaired.

17.7 HANDLING OF FIBER OPTIC CABLES

Fiber optic cables are made from a glass strand, which is covered with a polymer jacket. They are very thin, rigid, and have sharp ends. Handle a fiber optic cable with care during inspection, cleaning, and preparation of the fiber optic cable ends. Fiber optic cables should be cleaned using the cleaning material recommended by the manufacturer. Follow the recommended procedure for each fiber optic cable type during cleaning, handling, assembling, packaging, and storage. When cleaving a fiber optic cable, the loose scrap material is hard to see and can be very dangerous. Dispose of loose scrap immediately in a properly designated container. Do not touch the end of a stripped fiber optic cable. It easily penetrates skin and a fiber shard could break off. Do not rub your eyes when handling fiber optic cables; this would be extremely painful and requires immediate medical attention. Follow all safety procedures and regulations, and always wear the required personal protective safety equipment. Use safety goggles with side shields and wear protective rubber gloves or finger cots when handling fiber optic cables. Figure 17.13 shows different types of safety goggles. Always treat fiber optic cables as a potential hazard. Never look directly at the fiber optic cable ends during fiber optic assembly and testing.

17.8 EPOXY ADHESIVES AND SEALANTS

Epoxy adhesives and sealants are essential components in the manufacturing of electronic and optical fiber devices. There are different types and colors, depending upon the application. Epoxy adhesives come in several forms. One-part, two-part, and UV systems are the most common. A GRIN lens can be glued to a beamsplitter with an epoxy. Sealant materials are used in the packaging of electronic and optical fiber devices.

When using adhesives and sealant materials, be aware of their specifications. Specifications, applications, and handling procedures of these materials are found on the material safety data sheets (MSDSs). The MSDSs are available from the manufacturer, distributor, or may be downloaded from a number of websites. The adhesives and sealants are also very hazardous during storage, handling, and application. Prolonged or repeated exposure may cause eye or skin irritation. If contact does

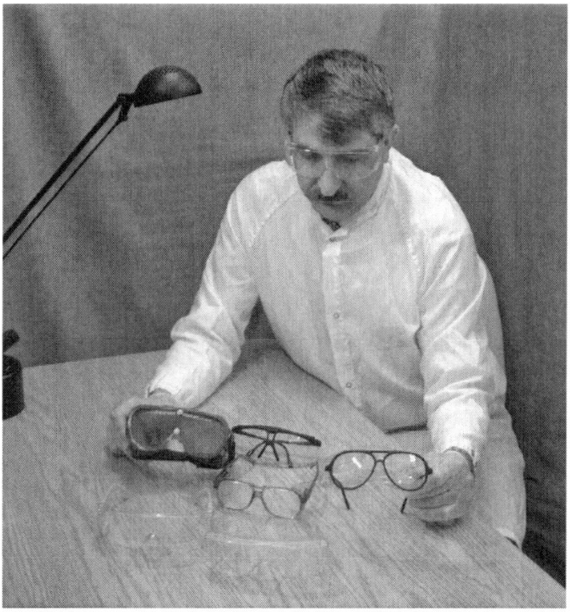

FIGURE 17.13 Safety goggles.

occur, wash the contact area immediately and seek medical help. Use safety goggles with side shields and wear protective rubber gloves or finger cots when handling adhesives and sealants. Follow all safety procedures and regulations, read the MSDS carefully, and wear the required personal protective safety equipment.

17.9 CLEANING OPTICAL COMPONENTS

Electronic and optical surfaces have to be clean and free of dust and other particles. Dust and other particles, however, can range in size from tenths to hundreds of microns in diameter. Their comparative size means that they can cover a part of the electronic and optical surfaces, and thus degrade the reflection or transmission quality of the data transmission in telecommunication systems. There are many standard procedures for cleaning electronic and optical surfaces. Before starting any cleaning procedure, locate the following standard equipment:

1. Cleaning material (denatured ethanol)
2. Cotton swabs
3. Tissue
4. Safety goggles
5. Finger cots or rubber gloves
6. Compressed air
7. Disposal container
8. Microscope with a magnification range about 50×

9. Infrared sensor card
10. Additional cleaning equipment:
    - Ultrasonic bath
    - Warm water and liquid soap
    - Premoistened cleaning wipe.
    - Polymer film

Some optical components (lenses, mirrors, prisms, beamsplitters, etc.) have special coatings, such as antireflection coatings, that are sensitive to solvents, grease, liquid, and mechanical abrasion. Take extra care and choose appropriate cleaning liquid and swabs when cleaning optical components with these coatings. The following is the preferred cleaning procedure for optical components:

- Wear rubber gloves and safety goggles.
- Hold a lens or a mirror by the rim, and a prism by the corners. Clean the optical component using a new, dry, or dampened swab with the recommended solvent. Rub the surfaces of the lens, using small circular movements, or one-directional movement on plane prism surfaces.
- Blow away any remaining lint with compressed air. This step depends on the optical component size and surface conditions. Check the air quality from the compressor before using it to clean optical components.

Some optical devices consisting of several optical components may not always be sealed completely. Therefore, use the recommended procedure to clean optical component surfaces without leaving any residue that could reduce the optical performance.

Disable all sources of power when cleaning any electronic and optical interface, such as the end of the ferrule on a fiber connector. Under any circumstances, do not look into the end of an optical device when the device is operating. Light from a laser device may not be visible to the human eye, but it can seriously damage the human eye.

17.10 OPTIC/OPTICAL FIBER DEVICES AND SYSTEMS

There has been a significant increase in the use of optic/optical fiber devices and systems in telecommunication networks. As optic/optical fiber devices become more common, it is important to understand the hazards associated with them. Optical fiber devices typically use a laser as a light source (sender and receiver). It is important to understand that not all lasers are created equal. Lasers are classified based on their output wavelength and power. Since they operate over a wide range of wavelengths and power outputs, the hazards arising from their use vary substantially.

Lasers are classified into four classes. Laser sources conformant to Class 1 and Class 2 do not cause serious damage, but eye protection must be taken into consideration. Class 3 and Class 4 lasers are powerful and can cause serious damage. Therefore, it is important to determine the class type of any optical equipment before working with it, assess the associated hazard, and comply with the safety requirements.

It is always a good practice to handle optical devices and measuring instruments with care. Normally, these devices and instruments are very expensive, sensitive, and may present a potential hazard if not used properly. Follow the recommended procedures for each device or instrument, to ensure proper handling during assembly, testing, packaging, and storage.

17.11 CLEANING CHEMICALS

Before the application of an epoxy or sealant, all surfaces should be treated using the recommended cleaning material. When using the cleaning materials, be aware of the appropriate precautions. Read all the information regarding cleaning materials in the MSDSs. All types of cleaning materials are a potential hazard; they may be flammable (even at low temperatures) and may pose other exposure risks. Use safety goggles with side shields and wear appropriate protective rubber gloves or finger cots. Follow all safety procedures and regulations. Use a ventilation hood when working with cleaning chemicals and epoxy adhesives, sealants, or any material that produces fumes.

17.12 WARNING LABELS

There are various types of warning labels used in buildings, transportation, services, industry, etc., to warn users about the level of danger ahead. Warning labels sometimes are called safety signs or safety messages. Safety signs clearly communicate by choosing the proper design and wording to suit safety needs. Standard signs, such as traffic warning signs and construction work labels, are available for general warnings.

Safety signs are divided into three general categories: danger, warning, and caution. They are also available in different sizes, colors, and with different graphics. Sometimes, a standard header can be used to create a new sign to suit a specific need. It is very important to use warning labels in manufacturing facilities to alert employees of any source of danger. These dangers may come from devices, instruments, chemicals, lasers, sounds, vibrations, biological hazards, etc. Employees should be introduced, in advance, to each source of danger in the manufacturing facilities and be shown the required personal protective safety equipment. Everybody must remember to consider safety first.

17.13 LASER SAFETY

A laser beam is a parallel, narrow, coherent, and powerful light source. It is increasingly powerful when concentrated by a lens. It is a hazard to human eyes and skin even at very low power.

All lasers are classified based on their potential power. These classifications are from the American National Standards Institute: ANSI Z136.1-1993, "American National Standard for Safe Use of Lasers," and Z136.3-1996, "American National Standard for Safe Use of Lasers in Health Care Facilities." Additional applicable standards include the Canada Labour Code and Canada Occupational Health and Safety Regulations (L-2-SOR/86-304).

These standards must be adhered to when using laser devices. These standards and codes are universally recognized as the definitive documents for establishing an institution, such as a school, factory, or hospital. Their basic classification system has been adopted by every major national and international standards board. These boards include the Center for Devices and Radiological Health (CDRH) in the United States and its Federal Laser Product Performance Standard, which governs the manufacture of lasers in the United States.

Lasers are classified into four laser classes: Class 1, 2 and 2a, 3a and 3b, and 4. Higher numbers reflect an increased potential to harm users. Figure 17.14 shows the laser warning labels, which are required to identify hazards from laser light sources.

The following criteria are used to classify the hazard level of lasers:

Wavelength: If the laser is designed to emit multiple wavelengths, the classification is based on the most hazardous wavelength.

Continuous Wave: For continuous wave (CW) or repetitively pulsed lasers, the average power output (Watts) and limiting exposure time inherent in the design are considered.

Pulse: For pulsed, lasers the total energy per pulse (joule), pulse duration, pulse repetition frequency, and emergent beam radiant exposure are considered.

Details of the laser classifications are listed below:

Class 1 lasers are laser devices with very low output power (between 0.04 and 0.40 mW), and operate in the lower part of the visible range (450 nm $< \lambda <$ 500 nm). These lasers are generally considered to be safe when viewed indirectly. Some examples of Class 1 laser devices include CD players, scanners, laser pointers, and small measurement equipment. Figure 17.15 shows the human eye. Laser light in the visible range entering the human eye is focused on the retina and causes damage to the eye. The most likely effect of intercepting a laser beam with the eye is a thermal burn, which destroys the retinal tissue. Never view any Class 1 laser beam directly.

Class 2 lasers are laser devices with low output power (must be less than 1 mW of visible CW), and operate in the visible range (400 nm $< \lambda <$ 700 nm). This class of laser could cause eye damage, if the laser beam is directly viewed for a very short period of time (more than 0.25 s). Some examples of Class 2 lasers are classroom demonstration laser sources and laser source devices for testing and telecommunications. Never view any Class 2 laser beam directly.

Class 2a lasers are low-output power devices, which are considered to be visible light lasers. This class of laser causes injury only when viewed directly for more than 1,000 s. This class must be designed so that intentional viewing of the laser beam is not anticipated. A supermarket bar code scanner is a typical example of a Class 2a laser device. Never view any Class 2a laser beam directly.

Class 3 lasers are divided into two subgroups (Class 3a and Class 3b lasers), so there is no plain Class 3.

Class 3a lasers are intermediate power devices. Class 3a lasers are allowed to exceed the output power limit of Class 2 by no more than a factor of 5, or visible light

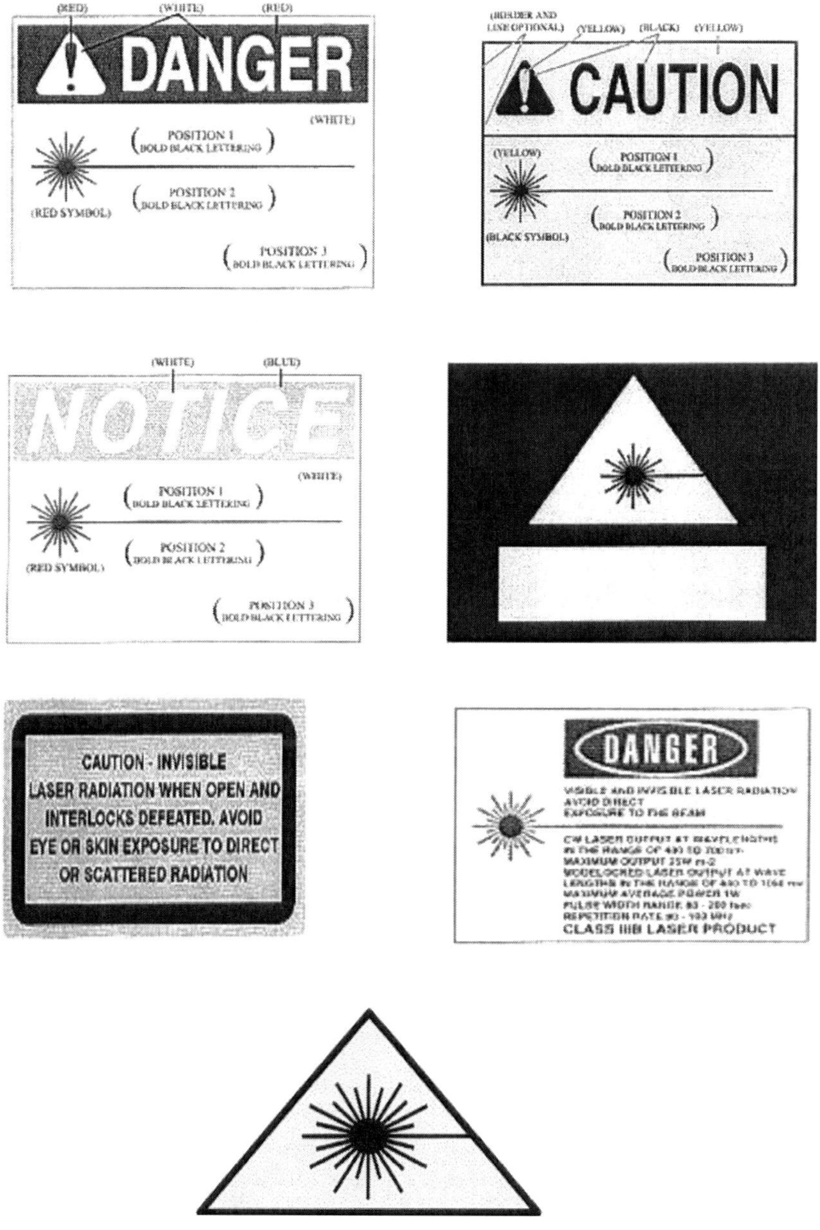

FIGURE 17.14 Laser warning labels.

power less than 5 mW. They are considered to be CW lasers. Often these lasers will have an expanded beam diameter so that no more than 1 mW can enter a fully dilated pupil, which is 7 mm. Some examples of Class 3a laser devices are laser scanners, laser printers, and laser source devices for testing and telecommunications. Direct viewing of a laser in this class could be hazardous to the eyes. Never view any Class

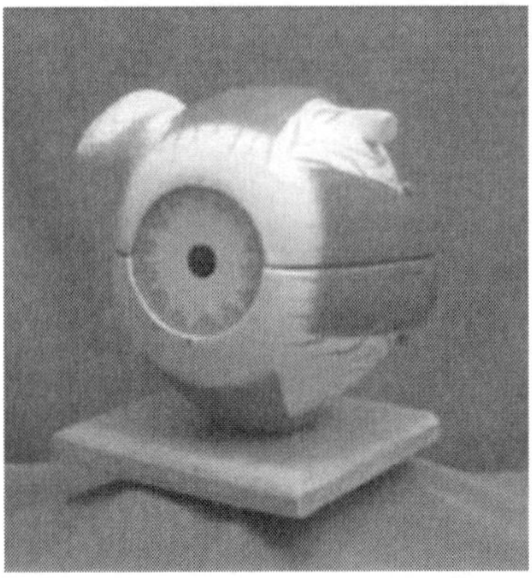

(a) Eye ball

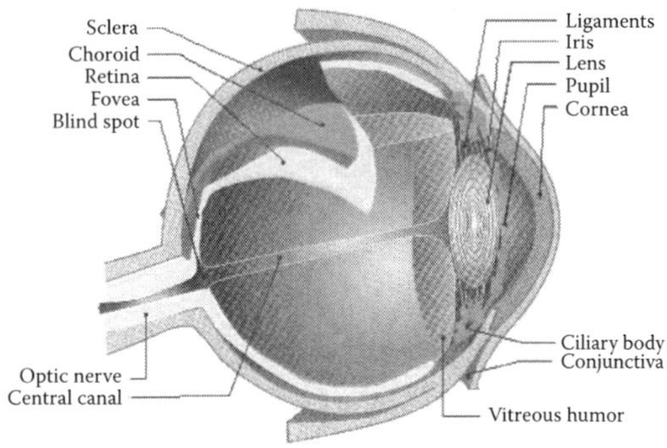

(b) Eye cross-section

FIGURE 17.15 The human eye.

3a laser beam directly. The laser beam wavelength may not be visible to the human eye, but causes damage to the eye and skin. Laser safety goggles for appropriate wavelength are required when working with this class.

Class 3b lasers are intermediate power devices; they output between 5 and 500 mW of CW, or else pulsed 10 J/cm^2 power. They are considered to be CW lasers. Scattered energy (diffuse reflection) is not considered hazardous in most situations, unless the laser source is operating near its upper power limit and the diffuse target is viewed at close range. Some examples of Class 3b lasers are laser source devices for testing. Never view any Class 3b laser beam directly or indirectly (any reflection by surrounding surfaces). The laser beam wavelength may not be visible to the human eye, but causes damage to the eye and skin immediately, with no time to react. Laser safety goggles for the appropriate wavelength are required when working with this class of laser.

Class 4 lasers are high-power devices; they output more than 500 mW of CW, or else pulsed 10 J/cm^2 power. They are considered to be very high-power lasers. Some applications of Class 4 laser devices include surgery, drilling, cutting, welding, and micromachining. When using Class 4 lasers, all types of reflections (whether direct, specular, and diffuse) are extremely hazardous to the eyes and skin. Class 4 laser devices can also be a fire hazard. Much greater control is required to ensure the safe operation of this type of laser device. Never view any Class 4 laser beam directly or indirectly (any reflection by surrounding surfaces). Be cautious of this type of laser. The laser beam wavelength may not be visible to the human eye but causes damage to the eye and skin immediately, with no time to react. Laser safety goggles for the appropriate wavelength are required when working with this class of laser.

Always follow all safety procedures, regulations, and wear the required appropriate personal protective safety equipment when using lasers. Never look directly or indirectly at a laser beam. Each institute should create appropriate safety procedures to guide employees and management in order to work in a safe environment. Each

TABLE 17.1
Institutional Program Requirements

Class	Power	Class Control Measures	Medical Surveillance	Safety and Training Program
1	No more than 0.04–0.40 mW	Not applicable	Not applicable	Not required
2	Less than 1 mW of visible, continuous wave light	Applicable	Not applicable	Recommended
2a	Less than 1 mW of visible, continuous wave light	Applicable	Not applicable	Recommended
3a	From 1 to 5 mW of continuous wave light	Applicable	Not applicable	Required
3b	From 5 to 500 mW of continuous wave light	Applicable	Applicable	Required
4	More than 500 mW of continuous wave light	Applicable	Applicable	Required

laser-designated area has to be controlled by a designated instructor/supervisor certified in laser safety. All laser safety requirements should be implemented in any laser-designated area. It is recommended to have an introduction course and workshop in laser safety for each laser classification.

According to the standard, knowing the classification of a particular device and comparing the information in Table 17.1 will usually eliminate the need to measure laser radiation or perform complex analyses of hazard potential.

17.14 LASER SAFETY TIPS

1. Do not enter the nominal hazard zone (NHZ). This zone is established according to the procedures described in ANSI Z136.1-1993. Enter this area accompanied with a designated instructor/supervisor certified in laser safety. Do not put any body part or clothing in the way of a laser beam.
2. Notice and comply with the signs and labels shown in Figure 17.14 that are posted on the laser-designated area door, devices, and equipment.
3. Wear the recommended eyewear and other protective equipment. Use laser safety goggles when you are in a laser-designated area, or in the vicinity, as shown in Figure 17.16.
4. Comply with the laser safety controls in the facility.
5. Attend laser safety training and workshops.
6. Update laser safety training and workshops as needed for employees and staff.
7. While assembling and operating laser devices, it is important to remember that laser beams can cause severe eye damage. Keep your head well above the horizontal plane of the laser beams at all times. Use white index cards to locate beamspots along the various optical paths.

FIGURE 17.16 Wear laser safety goggles.

8. When moving optical components, mirrors, or metal tools through the laser beams, the beam may reflect laser light momentarily at your partner or yourself. If there is a possibility of an accidental reflection during a particular step in an operation, then temporarily block or attenuate the laser beam until all optical components are in their proper place. It is a good policy to be aware of any stray laser beam reflections, and to warn your partner of any danger. If you are unsure of how to proceed safely with a given step in the operation of the laser device, ask the instructor/supervisor for assistance.

17.15 INDOOR AIR QUALITY

Concerns with indoor air quality (IAQ) have increased since energy conservation measures were instituted in office buildings during the 1970s, minimizing the infiltration of outside air and contributing to the buildup of indoor air contaminants. IAQ generally refers to the quality of the air in a work environment. Other terms related to IAQ include indoor environmental quality (IEQ) and sick building syndrome. Complaints about IAQ range from simple complaints, such as the air smelling odd, to more complex situations, where the air quality causes illness and lost work time. It may not be easy to identify a single reason for IAQ complaints because of the number and variety of possible sources, causes, and varying individual sensitivities.

IAQ problems can be caused by ventilation system deficiencies, overcrowding, tobacco smoke, microbiological contamination, outside air pollutants, and off-gassing from materials in the building and mechanical equipment. Related problems may also include comfort problems due to improper temperature and relative humidity conditions, poor lighting, and unacceptable noise levels, as well as adverse ergonomic conditions, and study-related psychosocial stressors. Typical symptoms may include headaches, unusual fatigue, itching or burning eyes, skin irritation, nasal congestion, dry or irritated throats, and nausea.

Ventilation is one of the most common engineering controls used to control emissions, exposures, and chemical hazards in the workplace. Other workplace environmental factors, including temperature, humidity, and odors, are also controlled with nonindustrial ventilation systems commonly known as heating, ventilating, and air conditioning (HVAC) systems.

- Management should have created guidelines for:
  1. Indoor air quality (IAQ)
  2. Building air quality (BAQ)
  3. Investigations, recommendations on sampling instrumentation and methods
  4. Management to prevent or alleviate IAQ problems and take acute health effects of major indoor air contaminants
- Management should have an overview of:
  1. Sources of indoor air pollution and health problems
  2. Ventilation, control, ventilation standards and building codes, and ventilation system problems
  3. Solutions for air cleaners and resolving problems

17.16 OTHER CONSIDERATIONS

These considerations apply to all employees, staff, and management:

1. Manufacturing activities, injuries, and illnesses are usually preventable by simply following the safely precautions in school throughout the year.
2. Never overload circuits, power bars, or connectors.
3. Lead innovative and cooperative efforts to improve manufacturing facilities safety, health, and the quality of employees' life.
4. Do not use or work with any device or equipment until it has been checked by qualified and authorized personnel in charge of the manufacturing facilities operation, as shown in Figure 17.17.
5. Everyone must wear personal protective equipment (safety goggles, protective gloves, ground connection, insulated tools, etc.) when working with electrical or laser equipment and chemicals.
6. Immediately report any damaged electrical or laser devices and equipment to the supervisor for immediate corrective action.
7. Managers should promote safety awareness among employees.
8. The nominal hazard zone (NHZ) should be established for each laser system.
9. Management should create and maintain a safe and healthy work and study environment.
10. Management and employees should understand the human and economic impact of poor safety and health in manufacturing facilities.

FIGURE 17.17 A certified technical staff member.

11. Management should create a safety checklist and maintenance and auditing programs for each manufacturing facility.
12. Eye protection should be worn at all times.
13. Eating, drinking, and smoking are not allowed in manufacturing facilities.
14. Unauthorized personnel should not be present in the manufacturing facilities or area, whether the lasers are operating or not.
15. Laboratory coats must be worn when handling cleaning, corrosive, toxic, or flammable materials. Gloves should be worn when necessary, especially when handling corrosive and highly toxic materials.
16. Never work alone in manufacturing facilities or workshops.
17. If a colleague is doing something dangerous, point it out to him or her and immediately inform the supervisor.
18. Know where safety equipment (eyewash, shower, extinguishers, emergency exits, first aid kit, etc.) is located and how to use it.
19. Know where the material safety data sheets (MSDSs) and Workplace Hazardous Materials Information System (WHMIS) are located and how to use them.
20. Know where the emergency phones and alarms are located and how to use them.
21. Know how to clean up chemical spills using the appropriate agents.
22. Preplanned experiments and a properly organized work area can eliminate a lot of potential safety problems. Clean up and decontamination must be a routine part of the experimental procedure for all employees.
23. Wash your hands after handling chemicals and before leaving the manufacturing facilities.
24. Ensure the manufacturing facilities safety program complements science.

Glossary

Absorption: The loss of light energy as it passes through a material. Loss is converted to other energy forms.

Acceptance angle: The maximum angle over which the core of a fiber optic cable accepts incoming light. The angle is measured from the centerline of the core.

Adapter: An adapter is a passive device used to connect together two different connector types.

Angle of incident (θ_i): The angle formed by an incident ray, and the normal line to the optical surface at the point of incident.

Angle of reflection (θ_{refl}): The angle formed by a reflected ray, and the normal line to the optical surface at the point of reflection.

Angle of refraction (θ_{refr}): The angle formed by a refracted ray, and the normal line to the optical surface at the point of penetration. The ray is refracted (bent) while passing from one transparent medium to another having different refractive indices.

Attenuation: The decrease in magnitude of the power of a signal in transmission media between points, normally measured in decibels (dB).

Backreflection: Reflection of light in the direction opposite to that in which the light was originally propagating.

Bandwidth: The measure of the information-carrying capacity of a fiber optic cable, normalized to a unit of MHz-km. This term is used to specify the capacity of a fiber cable.

Beam: A bundle of light rays that are diverging, converging, or parallel.

Beamsplitter: An optical device that divides incident light into two components (magnetic and electric).

Bend radius: The amount that a fiber optic cable can bend before the risk of breakage or increase in attenuation. The minimum bend radius is dependent on the fiber optic cable diameter and type.

Bidirectional device: An optical device that operates in both directions.

Birefringent: The separation of light beam, as it passes through a calcite crystal object, into two diverging beams, commonly known as ordinary and extraordinary beams.

Bundle of fiber cables: A group of fiber cables packaged together in a unit.

Candle or candela (cd): A unit of luminous intensity.

Cascade: An arrangement of devices, each of which feeds into the next.

Cladding: The layer of glass or other transparent material surrounding the light-carrying core of a fiber optic cable. The cladding has a lower refractive index than the core. The difference between the refractive indices creates an interface that confines light propagation in the core.

Coating: One or more thin layers of optical material applied to an optical surface to reduce reflection, create a mirror surface, absorb light, or protect the surface.

Connector: A device mounted on the end of a fiber optic cable that mates to a similar device to couple light into fiber cables. A connector joins a fiber cable end and a light source or detector in an optical device.

Core: The central part of a fiber optic cable, which conducts light. The core made from glass or plastic. It has a higher refractive index than the cladding.

Coupler: A device that connects two or more output ends, dividing one input between two or more outputs, or combining two or more inputs into one output. Couplers are used in telecommunication systems.

Coupling: Transfer of light into or out of a fiber optic cable. A coupling is used to launch a light source into a fiber optic cable.

Decibel (dB): The standard unit used to express loss. Decibel is defined as 10 times the base 10 logarithm of the ratio of the output signal to the input signal power.

Dispersion: The separation of a light beam into its various wavelength components. All transparent materials have different indices of refraction for light of different wavelengths.

Excess loss: Loss within a four-directional coupler. It is defined as 10 times the base 10 logarithm of the ratio of the signal power between ports 2 and 3, and port 1.

External auditing: A professional company is tasked to perform the external auditing activities for each field.

Ferrule: A part of a connector with a central hole that contains and aligns a stripped fiber cable in a connector assembly.

Fiber cable bundle: A group of fiber cables packaged together in a unit.

Fiber optic cable: An optical cable used for light transmission in telecommunications. Fiber optic cables come in a great variety of configurations.

Focal length (*f*): The distance between the second principal plane or equivalent refracting plane of a lens and the lens focal point when the lens is imaging an object at infinity. In a positive lens, the focal length is measured on the side of the lens opposite the object. In a negative lens, the focal length is measured on the same side as the object.

Fused fiber: Two fiber cables are heated, placed under tension, and caused to create a taper coupler join.

Fusion splice: A splice made by melting the ends of two fiber cables together by a spark shot so they form a permanent connection.

Graded-index fiber cable: A fiber optic cable whose core has a nonuniform index of refraction. The core comprises concentric rings of glass whose refractive indices decrease radially from the center of the core.

GRIN lens: An acronym for gradient index lens. They have a cylindrical shape. One end is polished at an angle of 2, 6, 8, or 12°. The other end is polished at an angle of 2 or 90°.

Hertz (Hz): The unit used to measure frequency. 1 Hz equals one wave or cycle per second.

Index of refraction (*n*): The ratio of the speed of light in a vacuum to the speed of light in a material.

Index-matching gel: A gel or fluid with a refractive index, which is matched to the refractive index of two fiber optic cores. It fills in the air gap between the fiber cable ends and reduces the Fresnel reflection that occurs in the gap.

Intensity: The light energy per unit area.

Interferometer: The interferometer, invented by American physicist A.A. Michelson (1852–1931), is an ingenious device that splits a light beam into two parts and then recombines them to form an interference pattern. The device can be used for obtaining accurate measurements of wavelengths, precise length measurements, and the accuracy of an optical surface.

Internal auditing: Internal auditing is completed by a group within the company.

ISO 9000: The ISO family has over 19,500 standards touching almost all aspects of daily life. The ISO 9000 family of quality management systems standards is designed to help companies ensure that they meet the needs of customers and other stakeholders while meetng statutory and regulatory requirements related to a product.

Jacket: A polymer (plastic, PVC, etc.) layer that covers the cladding/core layers of a fiber optic cable. The jacket has different colors and is used as a mechanical protection layer.

Kevlar: A strong synthetic material used in strength members in a fiber optic cable. A pull wire can be fastened to Kevlar during fiber optic cable installation.

LASER: An acronym for light amplification by the stimulated emission of radiation. Lasers produce the coherent source of light for fiber optic telecommunication systems.

Laser source: An instrument that produces monochromatic, coherent, collimated light.

Lens: One or more optical elements having flat or curved surfaces. If used to converge light rays, it is a positive lens; if used to diverge light rays, it is a negative lens. Usually made of optical glass but may be molded from transparent plastic. Lenses are sometimes made from a natural or synthetic crystalline substance to transmit very short wavelengths (UV) or very long wavelengths (IR).

Light: The form of electromagnetic radiation with a wavelength range of ~400 to ~700 nm. It generally travels in a straight line and exhibits the characteristics of both a wave and a particle.

Lightguide: A fiber optic cable or fiber optic cable bundle.

Light ray: The path of a single beam of light. In graphical ray tracing, a straight line represents the path along which the light travels.

Loss (dB): Attenuation of the power of a signal when it travels through an optical component. Normally measured in decibels (dB).

Lumen (lm): A SI unit of luminous flux. 1 lumen is the luminous flux emitted per unit solid angle by a light source having an intensity of 1 candela (cd).

Luminous intensity (I): Luminous intensity measures the brightness of a light source. The unit of measure is the candle or candela (cd).

Lux (lx): A unit of luminance equal to 1 lumen/m^2.

Mechanical splice: A process whereby two optical fibers are joined together using mechanical means.

Microbending: Tiny bends in a fiber optic cable, which allow light to leak out the core and introduce loss.

Micrometer (μm): One-millionth of a meter.

Microscope: An optical instrument used to inspect small objects that provide power magnification.

Mirror: An optical element with a smooth, highly polished surface (plane or curve) for reflective light. The reflecting surface is produced by a thin coating of gold, silver, or aluminum.

Mode: A term used to describe a light path passing through a fiber optic cable in single mode or multimode.

Multimode fiber: A fiber optic cable in which light travels in multiple modes.

Nanometer (nm): One-billionth of a meter. The unit usually used in specifying the wavelength of light.

Normal line: A reference line constructed perpendicular to an optical surface.

Numerical aperture (NA): The sine of half the angle in which the core of a fiber optic cable can accept or transmit light.

Object: The source figure in an optical system.

Optical axis: A straight line formed by joining the foci of lens.

Optical surface: The reflecting or refracting surface of an optical element.

Phase: The position of a wave in its oscillation cycle.

Photon: A particle or packet of radiant electromagnetic energy representing a quantum of light.

Photonics: The field of science and engineering encompassing the physical phenomena and associated with the generation, transmission, manipulation, detection, and utilization of light.

Plastic fiber cable: A fiber optic cable having a plastic core and plastic cladding. Plastic fiber cables are typically used in applications where sensitivity and loss are not important.

Plenum: The air space between walls, under structural floors, and above drop ceilings, which can be used to route interconnection cabling in a building.

Polarization: Alignment of the electric and magnetic fields that comprise an electromagnetic wave. If all light waves from a source have the same alignment, then the light is said to be polarized.

Prism: An optical element that is used to change the direction and orientation of a light beam. A prism has polished faces, which are used to transmit and reflect light.

Proximity sensor: A device that senses distance from a reflecting surface.

PVC: Polyvinylchloride. A material used in the manufacture of fiber optic cable jackets.

Ray: Straight lines that represent the path of a light ray.

Rayleigh scattering: The scattering of light, which results from small impurities in a material or composition.

Rectangular beamsplitters: Three prisms carefully cemented together along their hypotenuses. Polarization beamsplitters are used in optical devices where the output components are required to exit from the opposite side to the input signal. They also produce a lateral displacement between the two output components.

Reflection: The change in direction of a light ray when it bounces off of a reflecting surface.

Refraction: The bending of a light ray as it passes from one transparent medium to another of different refractive index.

Refractive index (*n*): The ratio of the speed of light in a vacuum to the speed of light in a specific material.

Ribbon fiber cables: Cables in which many optical fibers are embedded in a plastic flat ribbon-like structure.

Right-angle prism: A prism whose cross section is a right-angle triangle with two 45° interior angles. The prism faces, which are at right angles, are transmitting surfaces, while the hypotenuse face is a reflecting surface.

Scattering: Loss of light due to the presence of atoms in a transparent material.

Sensor: A device that responds to the presence of energy.

Sheath: An outer protective layer of a fiber optic cable. Fiber optic cables having an outer protective layer are suitable for indoor and outdoor cable installations.

Simplex fiber optic cable: A term sometimes used to describe a single fiber optic cable.

Single-mode fiber: A fiber optic cable in which the signal travels in one mode.

Snell's law: Describes the path that a light ray takes as it goes from one optical medium to another. It is also called law of refraction.

Speed of light: In vacuum, approximately 3×10^8 m/s.

Splice: A method for joining the ends of two fiber optic cables. There are two primary methods for splicing fiber optic cables: fusion and mechanical.

Strength member: That part of a fiber optic cable composed of Kevlar-armed yarn, steel strands, or fiberglass filaments. Strength members increase the tensile strength of a cable.

Switch: A device that regulates or directs a signal in telecommunication systems.

Total internal reflection: Total internal reflection of light occurs when light rays in a high-index medium exceed the critical angle (to the normal to a surface). This is the principal theory for explaining how light travels in the core of a fiber optic cable.

Visible light: Electromagnetic radiation, which is visible to the human eye. It has a wavelength range between 400 and 700 nm.

Waste: There are many types of waste in manufacturing lines.

Wave: One complete cycle of a signal with a fixed period.

Waveguide: A structure that guides an electromagnetic wave along its length. A fiber optic cable is an example of an optical waveguide.

Wavelength (λ): The period of a wave.

References

1. A. Bar-Lev, *Semiconductors and Electronic Devices*, 2nd ed., Prentice Hall, Upper Saddle River, NJ, 1984.
2. A. Berlin and K.J. Gabriel, Distributed MEMS: New Challenges for Computation, *IEEE Computational Science and Engineering Journal*, vol. 4, no. 1, pp. 12–16, 1997.
3. A.J. Jacobs-Cook, MEMS versus MOMS from a Systems Point of View, *Journal of Micromechanics and Microengineering*, vol. 6, pp. 148–156, 1996.
4. A. Marrakchi (ed.), *Photonic Switching and Interconnects*, Marcel Dekker, New York, 1994.
5. A.V. Uskov, et al., On Ultra-Fast Optical Switching Based on Quantum Dot Semiconductor Optical Amplifiers in Nonlinear Interferometers, *IEEE Photonics Technology Letters*, vol. 16, no. 5, pp. 1265–1267, 2004.
6. A. Yariv, Universal Relations for Coupling of Optical Power between Microresonators and Dielectric Waveguides, *Electronic Letters*, vol. 36, pp. 321–322, 2000.
7. A.P. Martin, Flawless Time Management, in *Practical Time-Saving Tips for Everyone*, Professional Development Institute, 1991, chap. 3.
8. Abdul Al-Azzawi, *Fiber Optics: Principles and Practices*, CRC Press, Taylor & Francis Group, Boca Raton, FL, 2006.
9. Abdul Al-Azzawi, *Light and Optics: Principles and Practices*, CRC Press, Taylor & Francis Group, Boca Raton, FL, 2006.
10. Abdul Al-Azzawi, Manufacturing Fiber Optic Connectors, lecture handout, Algonquin College, Ottawa, ON, Canada, 2011.
11. Abdul Al-Azzawi, Manufacturing Opto-Mechanical Switch, lecture handout, Algonquin College, Ottawa, ON, Canada, 2011.
12. Abdul Al-Azzawi, Manufacturing Photonics Component, lecture handout, Algonquin College, Ottawa, ON, Canada, 2011.
13. Abdul Al-Azzawi, Manufacturing Y-Coupler, lecture handout, Algonquin College, Ottawa, ON, Canada, 2012.
14. Abdul Al-Azzawi, *Photonics Principles and Practices*, CRC Press, Taylor & Francis Group, Boca Raton, FL, 2006.
15. Abdul Al-Azzawi, *Physical Optics: Principles and Practices*, CRC Press, Taylor & Francis Group, Boca Raton, FL, 2006.
16. Abdul R. Al-Azzawi and Peter Casey, *Fiber Optics Principles and Practices*, Algonquin College, Algonquin Publishing Centre, ON, Canada, 2002.
17. Adedeji B. Badiru, *Industry's Guide to ISO 9000*, John Wiley & Sons, New York, 1995.
18. S. Ariss, Employee Involvement to Improve Safety in the Workplace: An Ethical Imperative, *American Journal of Business*, vol. 18, no. 2, 2003.
19. Arthur Beiser, *Physics*, 5th ed., Addison-Wesley, Reading, MA, 1991.
20. B.E.A. Salah and M.C. Teich, *Fundamentals of Photonics*, John Wiley & Sons, New York, 1991.
21. B.J. Eggleton, et al., Microstructured Optical Fiber Devices, *Optics Express*, pp. 698–713, December 17, 2001.
22. Ken R. Beattie, Implementing ISO 9000: A Study of Its Benefits among Australian Organizations, *Total Quality Management*, vol. 10, p. 95, 1999.
23. Behzad Razavi, *Design of Integrated Circuits for Optical Communications*, McGraw-Hill, New York, 2003.

24. Bob Pease, When It Comes to Optical Fiber Installations, Some Companies Are Really, Blowing It, *Fiberoptic Product News*, p. 12, May 2004.
25. Breck Hitz, A New Tunable Optical Filter, *Photonics Spectra*, p. 32, October 2003.
26. Breck Hitz, Photonic Bandgap Fiber Eyed for Telecommunications, *Photonics Spectra*, pp. 122–124, April 2004.
27. Brian J. Carroll, Lean Performance ETP Project Management—Implementing the Virtual Supply Chain, STL St. Lucie Press, 2002.
28. Brian Rothery, *ISO 9000*, 2nd ed., Clays Ltd., St. Ives Place, UK, 1993.
29. Brian Rothery, *ISO 9000*, 2nd ed., Gower Press, Hampshire, UK, 1993.
30. Bruno Lotter, *Manufacturing Assembly Handbook*, Butterworths, Hartnoll Ltd., Cornwall, UK, 1989.
31. C. Camperi-Ginestet, et al., Micro-Opto-Mechanical Devices and Systems Using Epitaxial Lift Off, in *JPL, Proceedings of the Workshop on Microtechnologies and Applications to Space Systems*, June 1993, pp. 305–316.
32. C. Ray Asfahl, *Industrial Safety and Health Management*, 4th ed., Prentice Hall, Upper Saddle River, NJ, 1999.
33. C. Yeh, *Applied Photonics*, Academic Press, San Diego, 1994.
34. Canadian Health and Safety Legislation, *Ecolog Canadian Health and Safety Legislation*, Federal, Provincial, and Territorial Acts, Regulations, Guidelines, Codes, Objectives, Workers' Compensation, and WHMIS Legislation, 2000.
35. Chai Yeh, *Handbook of Fiber Optics: Theory and Applications*, Academic Press, San Diego, 1990.
36. Charles K. Kao, *Optical Fiber Systems: Technology, Design, and Applications*, McGraw-Hill, New York, 1982.
37. Cheryl Wischhover, *Elle*, July 9, 2013, http://www.elle.com/beauty/health-fitness/dangers-of-laser-hair-removal.
38. We-Minx Chow, *Assembly Line Design: Methodology and Applications*, Marcel Dekker, New York, 1990.
39. Christopher C. Davis, *Lasers and Electro-Optics, Fundamental and Engineering*, Cambridge University Press, New York, 1996.
40. Clyde N. Herrick and C. Lee McKim, *Telecommunication Wiring*, 2nd ed., Prentice Hall, Upper Saddle River, NJ, 1998.
41. Ruth Schwartz Cowan, *A Social History of American Technology*, Oxford University Press, New York, 1997.
42. D.C. Hood and M.A. Finkelstein, *Sensitivity to Light Handbook of Perception and Human Performance*, vol. 1, *Sensory Processes and Perception*, ed. K.R. Boff, L. Kaufman, and J.P. Thomas, John Wiley & Sons, Toronto, 1986.
43. D.L. Brower, et al., *Fizeau Interferometer for Measurement of Plasma Electron Current*, Electrical Engineering Department, University of California, Los Angeles.
44. D. Sadot and E. Boimovich, Tunable Optical Filters for Dense WDM Networks, *IEEE Communications Magazine*, vol. 36, pp. 50–55, 1998.
45. Dale H. Besterfield, *Quality Control*, 6th ed., OH, 2001.
46. Dale H. Besterfield, *Quality Control*, 7th ed., Pearson Education, Upper Saddle River, NJ, 2004.
47. Daniel H. Nichols, *Physics for Technology with Applications in Industrial Control Electronics*, Prentice Hall, Upper Saddle River, NJ, 2002.
48. Daniel Malacara, *Geometrical and Instrumental Optics*, Academic Press, Boston, 1988.
49. Data and Information Technology; Highways; Law; Safety and Human Factors; Society; I83: Accidents and the Human Factor, Files HSL, NTL, TRIS, USA, last modified May 14, 2009.

50. David Groth, et al., *Cabling—The Complete Guide to Network Wiring*, 2nd ed., Sams & Co., 2001.
51. David Loreggua, et al., Fizeau Interferometer for Global Astrometry in Space, *Applied Optics*, vol. 43, no. 4, pp. 721–728, 2004.
52. David Mazzarese, Meeting the OM-3 Challenge—Fabricating This New Generation of Multimode Fiber Requires Innovative State of the Art Designs and Testing Processes, *Cabling Systems*, pp. 18–20, December 2002.
53. David Page and Ian Routledge, Using Interferometer of Quality Monitoring, *Photonics Spectra*, pp. 147–153, November 2001.
54. David R. Goff and Kimberly S. Hansen, *Fiber Optic Reference Guide: A Practical Guide to the Technology*, 2nd ed., Butterworth-Heinemann, 1999.
55. De Camp, L. Sprague, and Catherine C. De Camp, *The Story of Science in America, Charles Scribner's Sons*, New York, 1967.
56. Dennis Derickson, *Fiber Optic Test and Measurement*, Prentice Hall, Upper Saddle River, NJ, 1998.
57. Djafar K. Mynbaev and Lowell L. Scheiner, *Fiber-Optic Communication Technology—Fiber Fabrication*, Prentice Hall, Upper Saddle River, NJ, 2005.
58. Donald J. Sterling Jr., *Premises Cabling*, Delmar Publishers, 1996.
59. Donald J. Sterling Jr., *Technician's Guide to Fiber Optics*, 2nd ed., Delmar Publishers, 1993.
60. E. Golovchenko, et al., Mutual Influence of the Parametric Effects and Stimulated Raman Scattering in Optical Fibers, *IEEE Journal of Quantum Electronics*, vol. 26, pp. 1815–1820, 1990.
61. E. Vail, et al., GaAs Micromachined Widely Tunable Fabry-Perot Filters, *Electronics Letters*, vol. 31, no. 3, pp. 228–229, 1995.
62. Eric Black, An Introduction to Pound-Drever-Hall Laser Frequency Stabilization, LIGO, California Institute of Technology, and Massachusetts Institute of Technology, 2000.
63. Eugene Hecht, *Optics*, 4th ed., Addison-Wesley Longman, Reading, MA, 2002.
64. F. Chan Him, Effects of Passivation of GaN Transistors, presented to Epitaxy Group of the IMS at NRC, Ottawa, Canada, 2005.
65. F.J. Duarte and J.A. Piper, Narrow-Linewidth, High Prf Copper Laser-Pumped Dye Laser Oscillators, *Applied Optics*, vol. 23, pp. 1391–1394, 1984.
66. F.J. Duarte, *Tunable Lasers Handbook*, Elsevier, 1999.
67. Francesca Cantore and Francesco G. Della Corte, 1.55-μm Silicon-Bases Reflection-Type Waveguide-Integrated Thermo-Optic Switch, *Optical Engineering*, vol. 42, no. 10, 2003.
68. M. Francon, *Optical Interferometry*, Academic Press, New York, 1966, pp. 97–99.
69. Frank J. Derfler Jr. and Les Freed, *How Networks Work*, millennium edition, Que Corporation, 2000.
70. T. Freiheit, Y. Koren, and S.J. Hu, Productivity of Parallel Production Lines with Unreliable Machines and Material Handling, *IEEE Transactions on Automation Science and Engineering*. (See also *IEEE Transactions on Robotics and Automation*, vol. 1, no. 1, 2004.)
71. G.H.B. Thompson, *Physics of Semiconductor Laser Device*, John Wiley & Sons, Chichester, UK, 1980.
72. G.K. Fedder and R.T. Howe, Multimode Digital Control of a Suspended Polysilicon Microstructure, *IEEE Journal on MEMS*, vol. 5, no. 4, pp. 283–297, 1996.
73. G.K. Fedder, S. Iyer, and T. Mukherjee, Automated Optimal Synthesis of Microresonators, in *Technical Digest of the IEEE International Conference on Solid-State Sensors and Actuators (Transducers '97)*, Chicago, June 16–19, 1997, vol. 2, pp. 1109–1112.

74. G.K. Fedder, S. Santhanam, et al., Laminated High-Aspect-Ratio Microstructures in a Conventional CMOS Process, *Sensors and Actuators A*, vol. 57, no. 2, pp. 103–110, 1996.
75. G.P. Agrawal and N.K. Dutta, *Semiconductor Lasers*, 2nd ed., Van Nostrand, New York, 1993.
76. B. Gebhard and C.P. Knowles, Design and Adjustment of a 20 cm Mach-Zehnder Interferometer, *Review of Scientific Instruments*, vol. 37, no. 1, pp. 12–15, 1966.
77. Gerald Vinten and Lawrence B. Sawyer, *The Manager and the Internal Auditor: Partners for Profit*, Wiley, New York, 1996.
78. Gerd Keiser, *Optical Communications Essentials*, 1st ed., McGraw-Hill, New York, 2003.
79. Gerd Keiser, *Optical Fiber Communications*, 3rd ed., McGraw-Hill, New York, 2000.
80. Gordon McComb, *The Laser Cookbook—88 Practical Projects*, Tab Book, Division of McGraw-Hill, New York, 1988.
81. Govind P. Agrawal, *Fiber-Optic Communication Systems*, 2nd ed., John Wiley & Sons, New York, 1997.
82. Grant Buckler, Air-Blown Fiber Facts, Focus on Engineering and Design, *Cabling Systems*, pp. 14–15, June/July 2003.
83. Grant Buckler, Fiber versus Copper, *CNS Magazine*, pp. 10–14, May/June 2004.
84. Gregory S. Romine, *Applied Physics Concepts into Practice*, Prentice Hall, Upper Saddle River, NJ, 2001.
85. L. Greiner, Evolution and Revolution as Organizations Grow, *Harvard Business Review*, pp. 37–45, July–August 1972.
86. H.H. Horng, et al., Designing Optical-Fiber Modulators by Using Magnetic Fluids, *Optics Letters*, vol. 30, no. 5, pp. 543–545, 2005.
87. H. Heidrich, et al., Passive Mode Converter with a Periodically Tilted InP/GaInAsP Rib Waveguide, *IEEE Photonics Technology Letters*, vol. 4, pp. 34–36, 1992.
88. P. Hariharan, Modified Mach-Zehnder Interferometer, *Applied Optics*, vol. 8, no. 9, pp. 1925–1926, 1969.
89. Harold Kolimbiris, *Fiber Optics Communications*, Prentice Hall, Upper Saddle River, NJ, 2004.
90. Harry J.R. Dutton, *Understanding Optical Communications*, IBM, Prentice Hall, Upper Saddle River, NJ, 1998.
91. Iñaki Heras, Gavin P.M. Dick, and Martí Casadesús, ISO 9000 Registration's Impact on Sales and Profitability—A Longitudinal Analysis of Performance Before and After Accreditation, *International Journal of Quality and Reliability Management*, vol. 19, no. 6, p. 774, 2002.
92. http://en. wiki pedia.org/wiki/Assembly, June 2012.
93. http://en. wikipedia.org/wiki/Just-in-Time, June 2012.
94. http://es.epa.gov/techinfo/case/michigan/mich-cs2.html, December 2012.
95. http://www.answers.com/topic/workshop#ixzz36AEgfjfe, July 2014.
96. http://www.apics.org/(10th Ed.), January 2012.
97. http://www.beyondlean.com/7-wastes, September 2013.
98. http://www.beyondlean.com/history-of-lean.html, October 2012.
99. http://www.borderwastewise.org/busassist/electr.htm, January 2012.
100. http://www.google.com/patents?hl=en&lr=&vid=USPAT4069764&id =HbEwAAAAEBAJ&oi=fnd&dq=types+of+waste+in+a+production+line, August 2011.
101. http://www.google.com/patents?hl=en&lr=&vid=USPAT4519999&id =pZQsAAAAEBAJ&oi=fnd&dq=types+of+waste+in+a+production+line, June 2008.
102. http://www.iee.org/oncomms/sector/management/magazine.cfm?issueID=144& articleID=4E4A7962-EAA3-3B43-45FCA290F78A527C, January 2010.

103. http://www.labour.gov.on.ca/english/hs/pubs/comp_erg/gl_comp_erg_6.php, October 19, 2013.
104. http://www.laserfx.com/BasicSafety/IndexBasicSafety.html, October 2013.
105. http://www.mle-consulting.com/mfgr/deadly_wastes.html, June 2012.
106. http://www.tpmonline.com/papakaizen/articls_on_lean_manufacturing_strategies/5s.htm, February 2012.
107. http://www.worksafenb.ca/docs/officeedist.pdf, October 19, 2013.
108. http://www.wsib.on.ca/files/Content/Fact%20Sheet_English5160A/5160A_Musculoskeletal_Disorders.pdf, October 20, 2013.
109. https://www.osha.gov/SLTC/etools/computerworkstations/checklist.html, October 19, 2013.
110. Intelligent Manufacturing, *Continuous Flow Manufacturing*, vol. 2, no. 12, 1996.
111. J.E. Vandemeer, M.S. Kranz, and G.K. Fedder, Nodal Simulation of Suspended MEMS with Multiple Degrees of Freedom, in *Proceedings of 1997 IMECE: The Winter Annual Meeting of ASME in the 8th Symposium on MEMS*, Dallas, TX, November 16–21, 1997.
112. J. Jackel, et al., 'Acousto-Optic Tunable Filters (AOTF's) for Multiwavelength Optical Cross-Connects: Crosstalk Considerations, *Journal of Lightwave Technology*, vol. 14, pp. 1056–1066, 1996.
113. J. Stone and L.W. Stulz, 'Pigtailed High Finesse Tunable FP Interferometer with Large, Medium and Small FSR,' *Electronics Letters*, vol. 23, pp. 781–783, 1987.
114. R.A. Jackson, The Laser as a Light Source for the Mach-Zehnder Interferometer, *Journal of Scientific Instruments*, vol. 42, pp. 282–283, 1965.
115. Jan K. Wachter, Ethics—The Absurd Yet Preferred Approach to Safety Management, *Professional Safety*, June 2011.
116. JDS Uniphase Corporation, Switch, SN Series, July 2003.
117. Jeff Hecht, *City of Light: The Story of Fiber Optics*, Oxford University Press, New York, 1999.
118. Jeff Hecht, *Understanding Fiber Optics*, 3rd ed., Prentice Hall, Upper Saddle River, NJ, 1999.
119. John Vacca, *The Cabling Handbook*, Prentice Hall, Upper Saddle River, NJ, 1998.
120. G. Johnson, G. Yip, and M. Hensmans, Achieving Successful Strategic Transformation, MIT *Sloan Management Review*, Spring, pp. 25–32, 2012.
121. R.D.M. Johnstone and W. Smith, A Design for a 6 in. Mach-Zehnder Interferometer, *Journal of Scientific Instruments*, vol. 42, pp. 231–235, 1965.
122. Joseph C. Palais, *Fiber Optic Communications*, 4th ed., Prentice Hall, Upper Saddle River, NJ, 1998.
123. Joseph Shamir, *Optical Systems and Processes*, SPIE Optical Engineering Press, 1999.
124. Jun Li, Mojtaba Kahrizi, and Leslie M. Landsberger, In-Plane Electro-Thermal Actuated Optical Switches, in *Opto-Canada, SPIE Regional Meeting on Optoelectronics, Photonics and Imaging*, vol. TD01, 2002.
125. K.J. Gabriel, O. Tabata, and S. Sugiyama, Surface-Normal Electrostatic Pneumatic Actuators, in Technical Proceedings of MEMS '92, Travemunde, Germany, February 1992, pp. 110–114.
126. K. Kuhn, *Laser Engineering*, Prentice Hall, Upper Saddle River, NJ, 1998.
127. K. Okamoto, *Fundamentals of Optical Waveguides*, Academic Press, San Diego, 2000.
128. Kenichi Hibino, Error-Compensating Phase Measuring Algorithms in a Fizeau Interferometer, *Optical Review*, vol. 6, no. 6, pp. 529–538, 1999.
129. Kenneth L. Arnold and Michael Holler, *Quality Assurance*, McGraw-Hill, New York, 1995.
130. Kevin P. Chen, In-Fiber Light Powers Active Fiber Optical Components, *Photonics Spectra*, pp. 78–90, April 2005.

131. Krause, T., The Ethics of Safety, *EHS Today*, 2007.
132. Lawrence B. Sawyer and Paul E. Heeschen, *Internal Auditor's Handbook*, Institute of Internal Auditors, 1984.
133. Lawrence B. Sawyer, *Elements of Management-Oriented Auditing*, Institute of Internal Auditors, June 1984.
134. Lawrence B. Sawyer, *Sawyer's Internal Auditing: The Practice of Modern Internal Auditing*, Institute of Internal Auditors, 1988.
135. Lawrence B. Sawyer, *The Manager and the Modern Internal Auditor: A Problem-Solving Partnership*, Institute of Internal Auditors, 1996.
136. L-Com Connectivity Products, Fiber Optic Connectors, master catalog, 2006.
137. Leo Setian, *Applications in Electro-Optics*, Prentice Hall, Upper Saddle River, NJ, 2002.
138. May-Fong Lim and I.A. Karimi, *Resource-Constrained Scheduling of Parallel Production Lines Using Asynchronous Slots*, Department of Chemical and Environmental Engineering, National University of Singapore, Singapore, Singapore.
139. Lin Zhang and Changxi Yang, Polarization Selective Coupling in the Holey Fiber with Asymmetric Dual Cores, LEOS2003, paper ThP1, Tucson, AZ, October 26–30, 2003.
140. Chris K.Y. Lo, Andy C.L. Yeung, and T.C. Edwin Cheng, Impact of ISO 9000 on Time-Based Performance: An Event Study, *World Academy of Science, Engineering and Technology*, vol. 30, no. 7, pp. 35–40, 2007.
141. M. Koga, Compact Quartzless Optical Quasi-Circulator, *Electronics Letters*, vol. 30, pp. 1438–1440, 1994.
142. M. Lequine, et al., Toward Tunable Thin-Film Filters for Wavelength Division Multiplexing Applications, *Applied Optics*, vol. 41, no. 16, pp. 3277–3284, 2002.
143. M. Mehregany and S. Roy, *Introduction to MEMS*, Microengineering Aerospace Systems, Aerospace Press, AIAA, El Segundo, CA, 1999.
144. M.S. Kranz and G.K. Fedder, Micromechanical Vibratory Rate Gyroscopes Fabricated in Conventional CMOS, in *Proceedings of the Symposium Gyro Technology*, Stuttgart, Germany, September 16–17, 1997.
145. Marion Cole, Telecommunications, Prentice Hall, Upper Saddle River, NJ, 1999.
146. Mark Ellwood, Stemming the Tide of Time at Work, Time Management, Canadian HR Report, December 29, 1997.
147. Mary Del Ciancio, Cognitive Robotics, *Manufacturing Autonation Magazine*, pp. 14–15, June 2007.
148. D. Miller, What Happens after Success: The Perils of Excellence, *Journal of Management Studies*, vol. 31, no. 3, pp. 325–358, 1994.
149. N. Sugimoto, et al., Waveguide Polarization-Independent Optical Circulator, *IEEE Photonics Technology Letters*, vol. 11, pp. 355–357, 1999.
150. Nabil Nahas, Mustapha Nourelfath, and Daoud Ait-Kadi, Selecting Machines and Buffers in Unreliable Series-Parallel Production Lines, Université Laval, Mechanical Engineering Department, Interuniversity Research Centre on Enterprise Networks, Logistics and Transportation (CIRRELT), Quebec, Canada, March 25, 2008.
151. Naga Shashidhar, Lensing Technology, *Fiber Product News*, pp. 14–15, April 2004.
152. C.A. Nanni and T.S. Alster, Laser-Assisted Hair Removal: Side Effects of Q-Switched Nd:YAG, Long-Pulsed Ruby, and Alexandrite Lasers, *Journal of the American Academy of Dermatology*, vol. 41, no. 2, pt. 1, pp. 165–171, 1999.
153. National Research Council, *Harnessing Light Optical Science and Engineering for the 21st Century*, National Research Council, National Material Advisory Board, National Academy Press, 1998.
154. NEL NTT Electronics, Thermo-Optic Switches, 2003–2004.
155. *Networks*, Prentice Hall, Upper Saddle River, NJ, 1993.
156. Newport Corporation, Photonics 2000 Catalog, Newport Corporation, 2000.

157. Newport Corporation, *Projects in Fiber Optics Applications Handbook*, Newport Corporation, 1986.
158. Nof. Shimon, W. Wilhelm, and H. Warnecke, *Industrial Assembly*, Chapman & Hall, New York, 1996.
159. O. Ishida, H. Takahashi, and Y.I. Noue, Digitally Tunable Optical Filters Using Arrayed-Waveguide Grating (AWG) Multiplexers and Optical Switches, *Journal of Lightwave Technology*, vol. 15, pp. 321–327, 1997.
160. OFS Leading Optical Innovations, Nonzero Dispersion Optical Fiber, True Wave Reach Fiber, product catalog, 2003.
161. OZ Optics, *Fiber Optic Components for Optoelectronic Packaging*, catalog, 2006.
162. Paula Tomkiel, Ethics of Workplace Safety, April 16, 2013.
163. *Photonics Buyers Guide*, PhotonicsBuyersGuide.com, 2014.
164. Bozena Poksinska, Jens Jörn Dahlgaard, and Marc Antoni, The State of ISO 9000 Certification: A Study of Swedish Organizations, *TQM Magazine*, vol. 14, no. 5, p. 297, 2002.
165. G. Probst and S. Raisch, Organizational Crisis: The Logic of Failure, *Academy of Management Executive*, vol. 19, no. 1, pp. 91–105, 2005.
166. R. Allen Shotwell, *An Introduction to Fiber Optics*, Prentice Hall, Upper Saddle River, NJ, 1997.
167. R.T. Bise, et al., Tunable Photonic Bandgap Fiber, in *Proceedings of Optical Fiber Communication Conference and Exhibit*, March 17, 2002, pp. 466–468.
168. Reza A. Maleki, *Flexible Manufacturing Systems—The Technology and Management*, Prentice Hall, Upper Saddle River, NJ, 1991.
169. Rita G. Lerner and George L. Trigg, *Encyclopedia of Physics*, 2nd ed., VCH Publishers, 1991.
170. Rob Walker and Neil Bessant, Image Fiber Delivers Vision of the Future, *Photonics Spectra*, pp. 96–97, March 2005.
171. Robert J. Hoss, *Fiber Optic Communications—Design Handbook*, Prentice Hall, Upper Saddle River, NJ, 1990.
172. Robert L. Wood, Ramaswamy Mahadevan, and Ed Hill, MEMS 2-D Matrix Switch, MEMS Business Unit, JDS Uniphase Corporation, Research Triangle Park, NC, 2000.
173. Robert W. Davies, *The Era of Global Transition—Crises and Opportunities in the New World*, Cass Business Press, City University London, UK.
174. Roberto R. Panepucci, *Integrated Micro Opto-Mechanical Systems (MOMS)*, NNUN Abstracts 2002/Mechanical and MEMS Devices, GalayOr Networks, 2002, p. 74.
175. Roger Allan, *3D MEMS-Based Optical Switch Handles 80-by-80 Fibers*, Electronic Design, Infineon Technologies, 2002.
176. S. Charschan, *Lasers in Industry*, Van Nostrand, New York, 1972.
177. S. He and R. Ben Mrad, A Vertical bi-directional Electrostatic Comb Driver for Optical MEMS Devices, in *Opto-Canada, SPIE Regional Meeting on Optoelectronics, Photonics and Imaging*, vol. TD01, 2002.
178. S.O. Kasap, *Optoelectronics and Photonics Principles and Practices*, Prentice Hall, Upper Saddle River, NJ, 2001.
179. Leticia Santos and Carmen Escanciano, Benefits of the ISO 9000:1994 System: Some Considerations to Reinforce Competitive Advantage, *International Journal of Quality and Reliability Management*, vol. 19, no. 3, pp. 321–44, 2002.
180. Scott Torrey, Gel-Free Outdoor Loose Tube and Ribbon Fiber Optic Cable, Corning Cable Systems, *Fiberoptic Product News*, pp. 14–16, August 2004.
181. Serge Ungar, *Fiber Optics: Theory and Applications*, John Wiley & Sons, New York, 1990.
182. Serope Kalpakjian and Steven R. Schmid, *Manufacturing Processes for Engineering Materials*, 5th ed., Pearson Education, Upper Saddle River, NJ, 2008.

183. Serope Kalpakjian and Stevev R. Schmid, *Manufacturing Engineering and Technology*, 6th ed., Prentice Hall, Upper Saddle River, NJ, 2010.
184. SETON, Signs, Labels, Tags, and Workplace Safety, catalog, 2006.
185. Stephanie Geist, Dmitry Gromov, and Jörg Raisch, Timed Discrete Event Control of Parallel Production Lines with Continuous Outputs, *Discrete Event Dynamic Systems*, vol. 18, no. 2, pp. 241–262, 2008.
186. Stephen C. Graves, Single Cycle Continuous Review Reducing Flow Time in Aircraft Manufacturing, February 1976, vol. 22, 629–638.
187. Stephen C. Graves, The Multiproduct Production Cycling Problem for Stochastic Science Intervals, *Management Science*, vol. 42, no. 1, pp. 1–18, 1996.
188. Sung-Sik Yun and Jong-Hyun Lee, A Micromachined In-Phase Tunable Optical Filter Using the Thermo-Optic Effect of Crystalline Silicon, *Journal of Micromechanics and Microengineering*, vol. 13, pp. 721–725, 2003.
189. T.N. Cornsweet, *Visual Perception*, Academic Press, New York, 1970.
190. Thomas P. Pearsall, *Photonics Essentials*, 2nd ed., McGraw-Hill, New York, 2010.
191. S. Tolansky, *An Introduction to Interferometry*, Longmans, Green and Co., London, 1955, pp. 115–116.
192. A. Topping, et al., A Review of the Ruby Laser with Reference to Hair Depilation, *Annals of Plastic Surgery*, vol. 44, no. 6, pp. 668–74, 2000.
193. Tunable Photonic Devices, *Optics Express*, pp. 1540–1550, April 19, 2004.
194. US Department of Labor, Workplace Safety and Health, retrieved from www.dol.gov, 2013.
195. Venkat Venkataramanan, *Introduction to Laser Safety*, Photonics Research Ontario, 2002.
196. Victor F. Weisskopf, How Light Interacts with Matter, *Scientific American*, pp. 60–71, September 1968.
197. Waldo T. Boyd, *Fiber Optics Communications, Experiments and Projects*, 1st ed., Howard W. Sams & Co., 1987.
198. Warren Hioki, *Telecommunications*, 3rd ed., Prentice Hall, Upper Saddle River, NJ, 1998.
199. D. White, Playing It Safe to Increase Profit, retrieved from www.sbnonline.com, 2006.
200. William K.Y. Tao and Richard R. Janis, *Mechanical and Electrical Systems in Buildings*, Prentice Hall, Upper Saddle River, NJ, 2001.
201. James R. Womack, Daniel Roos, and Daniel T. Jones, *The Machine That Changed the World: The Story of Lean Production*, Rawson Associates, New York, 1990.
202. www.advancedmanufacturing.com, February 2007.
203. www.buzzle.com, March 2012.
204. www.emsstrategies.com/dd030105articlel.html, September 2013.
205. www.industryforum.co.uk/glossary.htm, January 2012.
206. www.productivityeurope.org/main-content-pages/content/continuous-flow.html, October 2014.
207. www.themanager.org/Strategy/SCM.htm, February 2014.
208. www.transformsw.com/resources/automating.pdf, February 2014.
209. www.unidex-inc.com/lean.htm, May 2012.
210. www.wikipedia.org, March 2012.
211. Xinwan Li, et al., Digitally Tunable Optical Filter Based on DWDM Thin Film Filters and Semiconductor Optical Amplifiers, *Optics Express*, vol. 13, no. 4, 2005.

Index

D

E

K

L

M

N

O

P

Q

R

S